THE COMPLETE NAUTICAL
ASTRONOMER

The Complete Nautical Astronomer

CHARLES H. COTTER

EX.C., B.SC.(Lond.) M.SC.(Wales)
F.Inst.Nav.

*Senior Lecturer in the Department
of Maritime Studies at the
University of Wales Institute of
Science and Technology*

" They are ill discoverers who think
there is no land when they can see
nothing but sea."
FRANCIS BACON
The Advancement of Learning
Book 2.

AMERICAN ELSEVIER PUBLISHING COMPANY, INC.

NEW YORK 1969

American Edition Published by

AMERICAN ELSEVIER PUBLISHING COMPANY, INC.

52 Vanderbilt Avenue,

New York, N. Y. 10017

STANDARD BOOK NUMBER 444-19753-2

LIBRARY OF CONGRESS CATALOG CARD NUMBER: 70-78948

Printed and bound in Great Britain

CONTENTS

FOREWORD

Navigation is usually defined as that maritime art by which a mariner is able to get his vessel safely and expeditiously from one place to another.

The traditional methods available to the navigator for checking the progress of his vessel, once he is forced to leave the land astern, having embarked on an ocean voyage, are based upon the principles of mathematical astronomy. It is the astronomical methods of fixing a vessel, when shore observations or electronic aids to navigation are not available, with which we shall be concerned. This branch of navigation which is largely scientific in nature, being based upon mathematical principles, is known by the time-honoured name of *nautical astronomy*.

The *science* of nautical astronomy is definite and rigorous in its principles: it would not be a science otherwise. The rules may be taught by a teacher or learnt from a textbook, and laboriously applied in practice. On the other hand, the *art* of nautical astronomy (and it is interesting to note that in byegone days nautical astronomers were often styled artists) is, in contrast to the cold science, something warm and alive and peculiarly personal. A mariner's feeling for the subject, a feeling which is related to the degree of his wonder at the marvel of being able, in a wilderness of sea, to fix his vessel by means of the heavenly bodies, determines whether or not he is a true artist.

My aim in writing this book, which I regard as a companion volume to my *The Complete Coastal Navigator*, has been to provide a full and up-to-date text on nautical astronomy in the belief that a gap in the literature of navigation needs to be filled.

Now that systematic courses in navigation, of great educational as well as vocational value, are provided for students preparing for the examination leading to the Ordinary National Diploma in Nautical Science, and for those reading for the degree of B.Sc. in Nautical Science or Maritime Studies, the need exists for something more than mere manuals of navigation.

It is for this reason that I have, here and there, introduced a little historical background which I hope will lead to a broadening of the reader's knowledge of, and enhance his interest in, navigation.

It is my sincere hope that all who read this book, professional and amateur navigators alike—but especially students of the subject—will derive not only profit from its pages, but also that pleasure which springs from understanding and appreciating the remarkable and unfailing science which enables them to plan their landfalls with confidence and to complete their voyages safely.

Cardiff. August 1968   CHARLES H. COTTER

PART I

General Astronomy

CHAPTER I

The Universe

The world considered as a system, that is to say the aggregate of all existing things and space, assumed to be arranged in an orderly pattern, comprises the *universe*.

The universe, on account of its apparent orderly arrangement, is sometimes called the *Cosmos*; and the study of the science of the universe is, therefore, called *cosmology*. *Cosmogony* applies to any of a large number of theories of the creation of the universe; and *cosmography* denotes description of the universe.

The distribution of the stars throughout space is not uniform. Most of the stars visible from the Earth belong to a stellar system known as a *galaxy*. The Earth, which is a mere satellite of a star, belongs to a galaxy having a diameter of about 100,000 light years, and a maximum thickness of about 3000 light years. When the sky is observed on a dark, cloud-free and clear night along the plane of the local galaxy, the immense number of stars to be seen appears as a white belt girdling the heavens. This is known familiarly as the *Milky Way*.

For describing distances such as the dimensions of galaxies and the distances between stars, astronomers use (as we have done above) a practical unit of distance known as a *light year*. This is the distance traversed by light, which travels at the prodigious rate of 186,000 miles per second, in a year. The dimensions of galaxies are small compared with the distances which separate neighbouring galaxies: astronomical distances, therefore, have magnitudes wholly incomprehensible to most minds.

It appears that the principal 'physical' feature of the universe is space occupied by gaseous material at an infinitesimally low density. Interspersed throughout space are countless millions of stars.

A star comprises a vast quantity of material at a relatively high density. Nuclear processes within a star result in its temperature being exceedingly high. A star's high temperature renders it self-luminous by virtue of the enormous quantity of electro-magnetic energy, embracing a wide range of frequencies (including those of heat and light), which it radiates. To terrestrial dwellers the most prominent star is the Sun which, because of his proximity to the Earth—one of the Sun's family of planets —renders all other stars invisible when he is above an observer's horizon.

At night, when the Sun is below the horizon and the air is clear and cloudless, the stars in all their glory provide one of the most majestic and awe-inspiring sights of Nature. The study of the stars must have begun as soon as there were men to observe. The spectacle of the heavens is so wonderful that men could not have eyes to see and not fix them attentively thereupon.

The stars, because they lie in the expanse of the blue vault of heaven, are sometimes called *heavenly* or *celestial bodies*. The science which treats of the celestial bodies—their distribution, motions, sizes and constitutions—is called *astronomy*, from the Greek meaning the law of the stars or star distribution. In ancient times, practical astronomy—an art and science which served to provide the means of time-keeping, direction-measuring and calendar-making—was referred to as *astrology*. This science degenerated into the mere investigation of the aspects of the planets relative to one another and to the Sun, and their imagined influence on the destinies of men. Judging by the horoscopes printed regularly in some daily newspapers, this branch of knowledge still commands the attention of vast multitudes of modern men and women.

Astronomy is usually regarded as being the most ancient of the sciences. Folk at the earliest stage in human history were doubtless impelled by curiosity first to observe, and then to record, the movements of the celestial bodies. Having so many practical uses it is small wonder that astronomy was to become the first science to be cultivated by mankind. Investigation into the histories of every ancient people reveals their rude attempts to discover the laws governing astronomical phenomena such as eclipses of the Sun and Moon, and the seemingly curious motions of the planets.

The celestial bodies fall conveniently into two broad classes. One class comprises those which are relatively near to the Earth. These include the Sun, the planets (including the Earth) and the satellites of the planets. In addition to these are comets and meteors which frequently add splendour and interest to the night sky. Beyond the Solar System, at distances tremendously great compared with the dimensions of the Solar System, is the world of the stars. These form the other class of celestial bodies.

Observation of the heavens on a dark and clear night reveals a great diversity in the apparent brightness of the stars. A relatively small number are exceedingly bright. In contrast, the apparent brightness of each of a very large number is so close to the limit of visibility that it escapes notice unless it is viewed intently. Optical assistance, in the form of a telescope, reveals more and more stars in proportion to the magnifying power of the instrument. It would appear that space is boundless and the number of stars infinite.

The great distances that separate us from the stars (excepting the Sun) result in the real motions of the stars being optically unobservable except over relatively long periods of time. They are, therefore, called *fixed stars*. Observation of the night sky on successive nights manifests the unchanging pattern of the stars relative to one another.

THE CONSTELLATIONS

The grouping of the bright stars into *asterisms* or *constellations* originated in the mists of antiquity. The division of the stars into constellations is arbitrary. The names of most of the constellations stem from the Ancient Greeks who linked their fabulous history to the stars. Ptolemy of Alexandria, who flourished *c.* 150 AD, divided the stars visible at Alexandria into forty-eight constellations:

The Northern Constellations

1.	Ursa Minor	— The Little Bear
2.	Ursa Major	— The Great Bear
3.	Draco	— The Dragon
4.	Cepheus	— Cepheus
5.	Bootes	— The Herdsman
6.	Corona Borealis	— The Northern Crown
7.	Hercules	— Hercules
8.	Lyra	— The Harp
9.	Cygnus	— The Swan
10.	Cassiopeiae	— The Lady in her Chair
11.	Perseus	— Perseus
12.	Auriga	— The Waggoner
13.	Ophiuchus	— The Serpent-bearer
14.	Serpens	— The Serpent
15.	Sagitta	— The Arrow
16.	Aquila et Antinous	— The Eagle and Antinous
17.	Delphinus	— The Dolphin
18.	Equuleus	— The Horse's Head
19.	Pegasus	— The Flying Horse
20.	Andromeda	— Andromeda
21.	Triangulum	— The Triangle

The Zodiacal Constellations

22.	Aries	— The Ram
23.	Taurus	— The Bull
24.	Gemini	— The Twins
25.	Cancer	— The Crab
26.	Leo	— The Lion
27.	Virgo	— The Virgin
28.	Libra	— The Balance
29.	Scorpio	— The Scorpion
30.	Sagittarius	— The Archer
31.	Capricornus	— The Goat
32.	Aquarius	— The Water-bearer
33.	Pisces	— The Fishes

The Southern Constellations

34. Cetus	— The Whale
35. Orion	— Orion the Hunter
36. Eridanus	— Eridanus the River
37. Lepus	— The Hare
38. Canis Major	— The Great Dog
39. Canis Minor	— The Little Dog
40. Argo Navis	— The Ship Argo
41. Hydra	— The Hydra
42. Crater	— The Cup
43. Corvus	— The Crow
44. Centaurus	— The Centaur
45. Lupus	— The Wolf
46. Ara	— The Altar
47. Corona Australis	— The Southern Crown
48. Piscis Australis	— The Southern Fish

To these Hevelius, the wealthy astronomer of Dantzig, added another twelve constellations:

1. Antinous	— Antinous
2. Mons Menelai	— Mount Menelaus
3. Canes Venatici	— The Greyhounds
4. Camelolopardalis	— The Giraffe
5. Cerberus	— Cerberus
6. Coma Berenices	— Berenice's Hair
7. Lacerta	— The Lizard
8. Lynx	— The Lynx
9. Scutum Sobieskii	— Sobieskii's Shield
10. Sextans	— The Sextant
11. Triangulum Australa	— The Southern Triangle
12. Leo Minor	— The Little Lion

The famous Halley, a contemporary of Hevelius, after charting the southern skies added a further eight constellations:

1. Columba Noachi	— Noah's Dove
2. Robur Carolinum	— The Royal Oak
3. Grus	— The Crane

4. Phoenix   — The Phoenix
5. Pavo   — The Peacock
6. Apus   — The Bird of Paradise
7. Musca   — The Fly
8. Chamaeleon   — The Chameleon

The boundaries of the constellations were chosen arbitrarily, and they appeared differently in different star atlases. In recent times, a rectification of the boundaries has been made by the International Astronomical Union, the boundaries now consisting of arcs of great circles perpendicular to the equinoctial and arcs of small circles of declination.

Many of the brighter stars have particular names; but, in order to distinguish every star, it became necessary to adopt a simple and effective system other than by giving particular names. The astronomer Bayer introduced, in his famous star atlas *Uranometria* of 1603, a system in which each star in a constellation is assigned a letter of the Greek alphabet:

α alpha	ι iota	ρ rho
β beta	κ kappa	σ sigma
γ gamma	λ lambda	τ tau
δ delta	μ mu	υ upsilon
ϵ epsilon	ν nu	ϕ phi
ζ zeta	ξ xi	χ chi
η eta	o omicron	ψ psi
θ theta	π pi	ω omega

In general, the brightest star in a constellation is designated α, the next brightest β, and so on. Sirius, therefore, may be referred to as α Canis Majoris; and Rigel, the second brightest star in the constellation Orion, may be referred to as β Orionis. This general rule does not always apply. In particular it does not apply to the stars forming the Plough, which are lettered with the first seven stars of the Greek alphabet in order of position, not brightness, starting with Dubhe, which is α Ursae Majoris. When the Greek letters are exhausted recourse is made to Roman or italic letters in Bayer's system. Constellation numbers, instead of letters, were first suggested by **Flamsteed**, the first English Astronomer Royal, and this system is now almost universally used in all the great star catalogues.

The principal data contained in a star catalogue are the positions of the stars at a specified time or *epoch*, and the rates at which the positions are known or thought to be changing. From this information it is possible to compute the position of a star for any epoch.

The apparent brightness of a star depends upon its intrinsic brightness as well as upon its distance from the observer. Were all stars of equal intrinsic brightness it would be an easy matter to ascertain their relative distances, because the intensity of light received from a luminous source falls off as the square of the distance of the source from the observer. This however is not the case: there is a great range of intrinsic brilliance of stars; and the apparent brightness of a star gives no indication of its distance from an observer.

The Ancient Greek astronomers classified the stars into six categories of apparent brightness. Fourteen of the apparently brightest stars visible to them were designated stars of the *First Magnitude*. About fifty of the next brighter were designated stars of the *Second Magnitude*, and so on: the magnitude number increasing with decreasing brilliance. Stars just visible to the naked eye were designated stars of the *Sixth Magnitude*.

There is no sharp line of demarcation in the apparent brilliance of stars of consecutive magnitudes. And, moreover, because the estimation of stellar magnitudes depends upon optical comparison, it is impossible to state the magnitude of a given star with absolute numerical precision.

The Great Hipparchus, the ancient Prince of Astronomy, is sometimes said to be the first to have classified the stars according to their apparent brightnesses. Ptolemy of Alexandria improved on the rough classification invented by Hipparchus, and divided each class into three subdivisions. Stars having magnitudes between 2 and 4, for example, were classified as 'magnitude 3', 'magnitude 3 + ' or 'magnitude 3 − '.

The greatest improvement in the system of classification of stars according to apparent brightness came with the introduction of the decimal division of magnitudes. At the same time, the magnitude scale was extended, stars fainter than magnitude

6 having magnitude numbers greater than 6, and those brighter than magnitude 1 having decimal or negative magnitude numbers. This system was first used in the famous star catalogue prepared by Argelander and Schönfeld in the 19th century known as the *Bonn Durchmusterung* or B.D.

During the 19th century, careful observation and study revealed that the quantity of light corresponding to different magnitudes varied in geometrical progression from one magnitude to the next. This accords with a psycho-physiological law first enunciated by Fechner in 1859. Fechner's law states that the intensity of a sensation varies in arithmetical progression when the exciting cause varies in geometrical progression. If, therefore, the quantities of light received from two stars of magnitudes m_1 and m_2 are L_1 and L_2 respectively, then:

$$L_1/L_2 = k^{(m_2 - m_1)}$$

where k is a constant corresponding to the ratio of the brightnesses of two stars whose magnitude numbers differ by unity. This quantity is called the *light ratio*.

Sir John Herschel, the famous son of the illustrious Sir William Herschel, estimated that a star of magnitude 1 is 100 times as bright as a star of magnitude 6. It follows that if the ratio L_1/L_2 in the formula stated above is 100 and the index of k, that is $(m_2 - m_1)$, is 5, then:

$$100 = k^5$$

and

$$k = \sqrt[5]{100}$$

$$= 2 \cdot 51 \dots$$

$$= 2\tfrac{1}{2} \text{ approximately}$$

Thus:

A first-magnitude star is $2\tfrac{1}{2}$ times as bright as a second-magnitude star.

A second-magnitude star is $2\tfrac{1}{2}$ times as bright as a third-magnitude star.

A third-magnitude star is $2\tfrac{1}{2}$ times as bright as a fourth-magnitude star.

Also:

A first-magnitude star is $(2\frac{1}{2})^2$ times as bright as a third-magnitude star.

A first-magnitude star is $(2\frac{1}{2})^3$ times as bright as a fourth-magnitude star.

Etc.

Two stars are said to have the same magnitude when they appear to the eye to be of the same brightness. In measuring or comparing magnitudes visually, because of the uncertainty of human judgement, precise values or comparisons cannot be obtained. One difficulty in estimating relative brightnesses of stars arises from the diversity of colours of stars. It is difficult enough to compare the magnitudes of stars of the same colour, but when their colours are different, the difficulty is increased considerably. A source of uncertainty which influences an observer comparing the brightnesses of different coloured stars is called the *Purkinje phenomenon* after the physicist who first drew attention to it. Purkinje found that if the intensities of two lights of different colour having the same degree of brightness are changed equally, their relative brightnesses alter. It follows that two stars of different colour having the same magnitude when viewed with the naked eye will have different magnitudes when viewed through a telescope.

Photography plays an important role in modern astronomy. Photometric methods, of which there are several, are available for measuring the visual apparent magnitudes of stars. A photometric method involves comparing the brightness of a star with that of a standard light source or a standard selected star.

It may be thought that by photographing a star field the relative brightnesses of the images of the stars on the plate or film will be the same as that of the same stars viewed visually. This is not so because of the variety of colours of stars. The photographic plate or film is more sensitive to blue light and less sensitive to red light when compared with the human eye. It follows, therefore, that for a given exposure a blue star will produce a larger image on the plate or film than a red star of the same magnitude. The camera, therefore, cannot replace the human eye for determining apparent magnitudes of stars. Star photographs, however, when compared with visual observations are invaluable for estimating the degree of colour of stars. The difference between the 'photographic' magnitude and that observed visually is a function of the colour of the star called the *colour index*.

Photographic magnitudes depend upon the type of film or emulsion used, and are influenced by the use of coloured filters. When using photography for ascertaining and comparing star magnitudes standard films and exposures are used.

Atmospheric conditions often add to the difficulty of estimating star magnitudes. This follows because atmospheric absorption of light from stars varies with the clarity and humidity of that part of the atmosphere through which the light rays pass.

The apparent magnitude of Sirius, which is the brightest of the fixed stars, is -1.6. The second brightest star in the heavens is Canopus, which has a magnitude of -0.9.

There being no limit to the scale of magnitudes, an interesting problem of astronomy has been the measuring of the Sun's and Moon's magnitudes. The Sun's magnitude is reckoned to be about $-26\frac{1}{2}$, so that the light received from the Sun is about 10,000 million times as much as that received from the brightest fixed star. The magnitude of the Full Moon has been estimated to be about $-12\frac{1}{2}$.

The time and effort expended in becoming acquainted with the star groups leads to a form of pleasure and enjoyment which never loses its fascination. This is reason enough for anyone who wishes to become familiar with the patterns of the constellations. The nautical astronomer, however, has a more pressing reason for learning to recognize constellations so that he may readily identify the stars of navigation. *Navigational stars* are those for which astronomical data of use to the navigator are provided in his *Nautical Almanac*.

In the *Nautical Almanac* published jointly by the British and United States' Governments, astronomical data are provided for some 173 stars. These include all stars having a brightness of magnitude 3.0 or brighter. Of this number, fifty-seven are selected on the basis of brightness and distribution in the sky so as to give adequate coverage for normal navigational purposes. The fifty-seven *selected stars* are numbered according to their Sidereal Hour Angles. The number increases as the S.H.A. decreases. The selected stars have proper names in addition to their constellation designations. Their proper names are usually those by which they were known to the Arab or Greek astronomers of antiquity. The star charts provided in the *Nautical Almanac* serve admirably to show the distribution and celestial positions of all selected stars.

The expert nautical astronomer should have no difficulty in identifying any of the selected stars of navigation. The normal way of identifying a star, when the sky is cloudless, is from knowledge of the shape of the constellation of which it forms part. The magnitude and colour of a star may also assist in its identification. When the sky is partly clouded, identification of a star may require the use of a planisphere or star-finder, of which there are many varieties, or a star globe. Alternatively, a simple computation, using the star's altitude and azimuth and the observer's latitude, or a graphical solution by projection, may assist the navigator in solving the problem of star identification.

To learn the constellations requires frequent observation of the night sky. If someone who knows the sky is not available for assistance (and this would be most unlikely on board a ship) a star map will be of great help. Most seamen are able to identify a particular star by recognizing that it lies on a straight line or arc joining or passing through other known stars, or that it forms, with other known stars, a simple geometrical shape such as a triangle or a square. It is useful to be able to estimate the values of angles on the celestial sphere. With the aid of a star chart the following descriptions are designed to assist the novice in learning the night sky.

a. *Ursa Major and its Environs*

The constellation of Ursa Major, or the Great Bear, is so well known to northern observers that the seven stars of the Plough or Big Dipper provides a useful starting point for a survey of the constellations and the navigational stars.

The angle between Alkaid and Dubhe, the more northerly of the two stars (the other being Merak) which together form the so-called Pointers, is about 30°. Dubhe is about 30° from the celestial pole which lies on the straight line through the Pointers. At about the same angular distance—about 30°—from Alkaid, along an arc through the stars forming the handle of the Plough, the brilliant Arcturus will be found. Continuing along the arc through Alkaid and Arcturus, and again at about 30° from Arcturus, will be found the bright star Spica, associated with which will be found the four stars of the constellation Corvus, forming what seamen call Spica's Spanker. Continuing along the same arc through Alkaid, Arcturus, and Spica, and again at about 30° from Spica, will be found Alphard, the 'lonely' star, so named because it lies in a region in which bright stars are scarce.

On the concave side of the arc through Alkaid, Arcturus, Spica, and Alphard, the beautiful constellation of the Lion will be found, with Regulus and Denebola, the former star being located at the handle of the Sickle. On the convex side of the arc, lying between Alkaid and Arcturus, is Alphecca of the Northern Crown, which star, together with Alkaid and Arcturus, forms an isosceles triangle with the base having an angular distance of about 15°.

What is perhaps the most important navigational star, providing, as it does, a ready means of finding latitude, is Polaris, the brightest star in the constellation Ursa Minor. Polaris is easily located if the Pointers of the Plough are visible. Moreover, being so close to the north celestial pole, its altitude is approximately equal to the northern latitude of an observer, and its bearing is approximately due north.

At the end of a great-circle arc through Polaris, with Polaris at about the central position and the other end located in the vicinity of the Plough, will be found the constellation of Cassiopeiae, the Lady in the Chair. The five brightest stars in this constellation form the shape of a letter W, with Schedar standing at the right foot below Caph (β Cassiopeiae) which lies at the top of the right-hand side of the W. The straight line from Caph through Schedar fetches up at Algol (α Cephei) at an angular distance of about 20° from Schedar.

b. Orion and its Environs

The constellation of Orion the Hunter is often regarded as being the most splendid of all the star groups.

To the Arab astronomers of old, the constellation Orion was called the Central One, because it stands astride the celestial equator. The four principal stars of Orion form the shoulders and feet of the Hunter. From the brightest star, the brilliant red Betelgeuse located in the north-east shoulder of Orion, to the second brightest blue-white Rigel, which forms his south-east foot, the angular distance is about 20°. The three stars forming Orion's belt stand at the mid position of a straight line which terminates at the brilliant white Sirius, the Dog-star, which lies about 20° to the south-west of the belt, and the ruddy Aldebaran, or Bull's Eye, which lies about 20° to the north-east of the belt.

Due north of the three stars of Orion's belt, and passing through a tiny group of faint stars forming the Hunter's head, lies the star Elnath, located about 20° north of Orion's shoulders. A further 20° along this line lies the brilliant pale yellow Capella or the Goat, in the constellation Auriga the Goatherd. Capella is readily recognized by the three small stars, known as the Kids, which form a small isosceles triangle, and which accompany her.

To the north-west of Betelguese, at a distance of about 25°, are the two bright stars Castor and Pollux of the constellation Gemini the Twins. Castor, or α Geminorum, is the more northerly and brighter of the pair. It is a brilliant white star compared with Pollux which is yellow. Due south of the Heavenly Twins, and due west of Betelguese is Procyon, another beautiful yellow star of magnitude −0·5. Procyon lies at the eastern end of a straight line through Regulus, Denebola and Arcturus.

Due east of Orion's right shoulder, at which the star Bellatrix is located, at an angular distance of about 30°, is Menkar, the brightest star in the constellation Cetus the Whale. Due south of Orion's feet, at about 25° from Rigel, is Phact, the brightest star in the constellation Columba the Dove.

c.   Pegasus and its Environs

The three stars Markab (α Pegasi), Scheat (β Pegasi), and Algenib (γ Pegasi), which belong to the constellation of the Flying Horse, together with Alpheratz the brightest star of the constellation Andromeda, are located respectively at the corners of a conspicuous square of side about 15°. On a straight line extending to the north-west of Alpheratz are located Mirach (β Andromedae) and Almach (γ Andromedae). This line continues to Mirfak, the brightest star in the constellation Perseus. The adjacent stars in this set of four are about 10° apart. Mirfak lies about midway between Cassiopeiae and Capella.

To the north-east of Scheat (β Pegasi), at a distance of about 20°, is the striking constellation Cygnus, the Swan. The five brightest stars in this constellation form a cross set in the midst of the Milky Way. The star at the head of the cross is Deneb, which name, meaning tail, stands at the tail of the Swan, the head star of which corresponds with the foot of the cross. This is Albireo, a beautiful double star, one gold and the other blue, whose angular separation is about half a minute of arc.

Lying between Alpheratz and Algenib, at a distance of about 20°, are the two stars Hamal and Sheratan, the two brightest stars of the constellation Aries the Ram.

To the south-east of Markab, at about 12° from it, is Enif. About 20° to the east of Enif is the conspicuous star Altair, the brightest star in the constellation Aquila, the Eagle. Altair, a first-magnitude star, is recognized in that it lies midway between two stars of the third or fourth magnitude, each about 3° from it. These three stars provide a pointer to the beautiful white star Vega, the brightest star in the northern skies. Vega lies about 15° to the east of the constellation Cygnus.

At about 45° due south of Markab is the bright star Fomalhaut, the principal star in the constellation Piscis Australis.

d.   The Southern Stars

Perhaps the most notable constellation of the southern skies is the Southern Cross, which lies about 30° due south of Spica's Spanker. The brightest star of the Southern Cross is named

Acrux. This is the most southerly of the four stars which form the cross. The star known as Gacrux stands at the north end of the upright part of the cross about 6° from Acrux.

At about 10° to the west of the Southern Cross lie the two brightest stars of the constellation Centaur, the brighter of the two being Rigil Kentaurus which is the more westerly.

At about 35° to the south of Sirius, and about 30° to the east of the Southern Cross, is the second brightest star of the heavens. This is Canopus, the brightest star in the constellation Argo. To the east of, at about 30° from, Canopus is the brilliant Archernar standing on one end of the heavenly river Eridanus, the other end of which is located near Rigel in Orion, and which is marked by the second brightest star in the constellation Eridanus—a star which forms the end of Orion's sword.

CHAPTER II

The Solar System

The Solar System comprises the Sun and the planets which revolve around him, and whose orbital motions are controlled by the Sun's gravitational force. The planets, in order of distance from the Sun, are illustrated in Fig. 1.

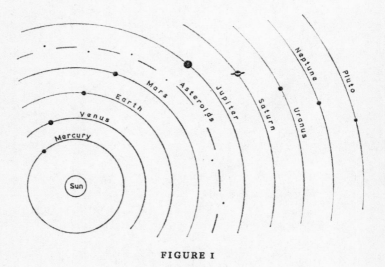

FIGURE I

Several of the nine planets have satellites: the Earth has one, but Jupiter, which is the largest of the planets, has at least nine. The satellite or satellites of a planet revolve around the parent planet in much the same way as the planets revolve around the Sun. In addition to the planets and their attendant satellites other members of the Solar System are the *Minor Planets* or

Asteroids the orbits of which lie mainly between those of Mars and Jupiter. *Comets* and *meteors* or 'shooting stars' also belong to the Solar System.

The Sun, which lies at the central position of the Solar System, rotates about a diameter with a period of about $24\frac{1}{2}$ days. The planets and their satellites also rotate: the Earth's period of rotation being a day. The directions of rotation of the Sun, planets and their satellites are, with minor exceptions, the same as that of the motions of the planets around the Sun and the satellites around their parent planets.

The Sun, whose diameter is about 865,000 miles, has a mass equal to about 750 times the mass of the remainder of the Solar System combined.

The path of a planet around the Sun is known as the *orbit* of the planet. The plane of the orbit of a planet tends to be fixed in space, but the orbits of the planets are not co-planar, although they are nearly so.

The orbit of a planet is an ellipse having the Sun at one of its focal points. The time taken for a planet to make one revolution around the Sun is known as the planet's *period of revolution*. The Sun's gravitational force on a planet varies directly as the mass of the planet and inversely as the square of the distance between the planet and the Sun. The nearer planets, therefore, regardless of their masses, have smaller periods than the more remote ones. The following table summarizes the more important astronomical data of the planets.

The planets are not, like the Sun, self-luminous: they are

NAME	SYMBOL	AVERAGE RADIUS OF ORBIT	PERIOD	ORBITAL VELOCITY
Mercury	☿	$36\cdot0 \times 10^6$ ml	88 days	$30\cdot0$ m.p.sec
Venus	♀	$67\cdot2 \times 10^6$	225 ,,	$22\cdot0$,,
Earth	⊕	$92\cdot9 \times 10^6$	365 ,,	$18\cdot6$,,
Mars	♂	$141\cdot5 \times 10^6$	687 ,,	$15\cdot0$,,
Jupiter	♃	$483\cdot5 \times 10^6$	12 years	$8\cdot0$,,
Saturn	♄	$886\cdot5 \times 10^6$	30 ,,	$5\cdot0$,,
Uranus	♅	$1782\cdot0 \times 10^6$	84 ,,	$4\cdot2$,,
Neptune	♆	$2792\cdot0 \times 10^6$	164 ,,	$3\cdot4$,,
Pluto	♇	$3716\cdot0 \times 10^6$	248 ,,	? ,,

rendered visible by reflected sunlight. Because of their proximity to the Earth, compared with the remote fixed stars, they move comparatively rapidly relative to the stars. It is for this reason that they are known as planets—the word planet being derived from the Greek word meaning wanderer. The planets Uranus, Neptune and Pluto were discovered in comparatively recent times: the other planets, therefore, are often referred to as the *ancient planets*. The planets Venus, Mars, Jupiter and Saturn are, at times, suitable for navigational purposes. They are, therefore, called the *navigational planets*, and astronomical data related to them are tabulated in the seaman's *Nautical Almanac*.

One of the significant discoveries in the history of astronomy was that made by the famous astronomer Johannes Kepler (1571–1630). Kepler, after extended observation of the planet Mars, and comparison of his observations with those made by his illustrious master Tycho Brahe (1546–1601), demonstrated that the orbit of Mars around the Sun is elliptical, having the Sun at one of the focal points of the ellipse. He also found that the orbital velocity of Mars varied during the time of revolution, being greatest when it was nearest to, and least when farthest away from, the Sun. Similar study of the apparent motions of the other planets revealed that their orbits also are elliptical. From these observations Kepler formulated his famous laws of planetary motion:

1. Every planet revolves around the Sun in an elliptical orbit having the Sun at one focus of the ellipse.
2. The line joining a planet to the Sun sweeps out equal areas in equal time intervals.

A third law relates the period of a planet's revolution and its mean distance from the Sun:

3. The square of the period of a planet's revolution around the Sun is proportional to the cube of its mean distance from the Sun.

In other words, if the period is T and the mean distance from the Sun is D, then:

$$T^2 \propto D^3$$

and
$$T^2 = kD^3$$

where k is a constant.

It follows, therefore, that if the Earth's period and mean distance from the Sun are known, the period and orbits of the other planets may be found.

It is interesting to note that Kepler formulated his laws empirically, and that it was not until after his death that Kepler's laws were demonstrated mathematically by Sir Isaac Newton (1642–1727) who showed that they were consequential to the law of universal gravitation.

The point in a planet's orbit which is nearest to the Sun is called *perihelion*, and the most remote point is called *aphelion*. The Earth is at perihelion at about January 3rd and at aphelion at about July 3rd each year.

The orbital motions of the planetary satellites conform to Kepler's laws. The point in the Moon's orbit nearest to the Earth is called *perigee*, and the most remote point is called *apogee*.

The two planets Mercury and Venus are known as *inferior planets* because their orbits lie within the Earth's orbit. The other planets, whose orbits lie outside the Earth's, are known as *superior planets*.

When a planet and the Sun lie in the same direction from the Earth, the planet is said to be *in conjunction* with the Sun. When a planet lies in a direction opposite to that of the Sun it is said to be *in opposition* to the Sun. When the angle at the Earth between a planet and the Sun is 90°, the planet is said to be *in quadrature* with the Sun.

An inferior planet can never be in opposition or quadrature with the Sun. This will readily be seen from Fig. 2.

In Fig. 2, the Earth is assumed to be at E. When Mars (or other superior planet) is at M_1, it is in conjunction with the Sun. When at M_2 it is in opposition, and when at M_3 or M_4 it is in quadrature with the Sun.

When Venus (or other inferior planet) is at V_1 it is said to be at *inferior* conjunction, and when at V_2 it is said to be at *superior* conjunction. The angle at the Earth at any instant between the directions of the Sun and any planet is called the *angle of elongation* of the planet at the instant. When a planet is in conjunction the angle of elongation is 0°. When it is at quadrature the angle of elongation is 90° and when it is in opposition it is 180°. It will be seen from Fig. 2 that when Venus is at V_3 or V_4 such that EV_3 and EV_4 are tangents from the Earth to its orbit, Venus will be at maximum angle of elongation, which is about 47°.

When a planet lies to the west of the Sun it will normally set before sunset and rise before sunrise. In this circumstance it will be visible for some period of time before sunrise. This

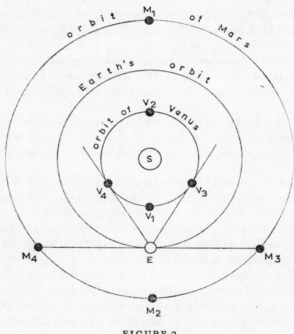

FIGURE 2

will be so when the planet has westerly elongation. When a planet lies to the east of the Sun it has easterly elongation and will rise after sunrise and set after sunset. It is for these reasons that a planet having westerly elongation is called a *morning star*, and one having easterly elongation, an *evening star*.

The apparent path of a planet as observed from the Earth is the resultant of its own orbital motion and that of the Earth's around the Sun. The planets trace out complex paths often forming a series of successive loops known as *retrogressive loops*. When an inferior planet is increasing its westerly elongation following inferior conjunction, it moves westwards relative to the fixed stars; but, when it is increasing its easterly elongation after superior conjunction, it moves eastwards relative to the fixed stars. When moving eastwards the apparent motion of the planet is said to be *direct*. When moving westwards its motion is said to be *retrograde*. The middle of the period of retrogression of an inferior planet takes place at inferior conjunction.

When an inferior planet is at inferior conjunction its illumin-ated hemisphere is directed away from the Earth. When at superior conjunction it appears as a disc of light—its illumin-ated hemisphere then being directed towards the Earth. Because the proportion of the illuminated hemisphere varies with time— the proportion depending upon the angle of elongation and whether it is increasing or decreasing—inferior planets exhibit phases as does the Moon.

THE EARTH

For most purposes of navigation the Earth may be considered to be perfectly spherical.

The Earth, in common with all spinning bodies, possesses the property of gyroscopic inertia by virtue of which her plane and axis of rotation tend to be maintained.

As a result of the Earth's slow, but uniform, rate of spin, during the course of a day—which is the period of the Earth's rotation—every celestial object rises out of the eastern half of an observer's horizon, and sets into the western half, crossing the north–south vertical plane through the observer at the middle time of those of its rising and setting.

The real motions of the stars, because stars are so far distant from the Earth, are not readily observable. It is for this reason that they are said to be fixed. The stars, and indeed all other celestial bodies, including the Sun, Moon and planets, are often assumed to lie on the inside of a sphere of infinite radius known as the *celestial sphere* which is centred at the Earth or Sun. The rotation of the Earth about her polar axis is manifested by an apparent diurnal revolution of the celestial sphere about an axis the ends of which are called the *north* and *south celestial poles*.

Our fundamental ideas of horizontal direction are related to the Earth's rotation. Every point on the Earth's surface is continually being carried in a direction called *east*. The horizontal direction opposite to east is *west*. The directions *north* and *south* are 90° to the left and right, respectively, of east. The natural compass of an observer located on the Earth is his *celestial horizon*—the great circle on the celestial sphere which divides the heavens into the visible and invisible hemispheres.

The axis of the apparent diurnal revolution of the celestial sphere towards the west is coincident with the axis of the Earth's spin towards the east.

The great circle on the Earth which lies in the plane of the Earth's spin is called the *equator*. The equator divides the Earth into the northern and southern hemispheres. The axis of the spin of the Earth terminates at the Earth's poles—the North Pole in the northern hemisphere and the South Pole in the southern hemisphere.

Small circles parallel to the equator are called *parallels of latitude*. Semi-great circles which extend from pole to pole across the equator and every parallel of latitude at right angles. These are called *meridians*.

To define a terrestrial point we state two angles or their corresponding spherical distances—one related to the parallel of latitude, and the other to the meridian which intersects the parallel at the point. These angles are called *latitude* and *longitude*.

Treating the Earth as a sphere, the latitude of a place is the angle at the Earth's centre measured in the plane of the meridian of the place, between the radii which terminate at the place and the equator respectively. The latitude of any place in the northern hemisphere is named *north*, and that of any place in the southern hemisphere is named *south*.

The plane of reference from which latitudes are measured is that of the equator, the latitude of every point on which is 0°. The latitude of either pole is 90°.

The longitude of a terrestrial position defines the meridian on which the position lies. The meridian from which longitudes are measured is chosen arbitrarily, and is called the *prime meridian*. The prime meridian is sometimes called the *Greenwich meridian*, because it is the meridian on which the transit instrument at the original Greenwich observatory rested.

The word meridian is derived from the circumstance that when the Sun, in his apparent diurnal path across the sky, crosses the meridian of the place, it is midday at the place.

The prime meridian and the antipodal meridian, which are separated by 180° of longitude, divide the Earth into the *eastern* and *western hemispheres*. All places which lie to the east of the prime meridian and to the west of the 180th meridian are in the eastern hemisphere and their longitudes are named *east*. All other places are in the western hemisphere and their longitudes are named *west*.

The longitude of a terrestrial position is the smaller angle at the pole or the smaller arch of the equator contained between the prime meridian and the meridian of the place.

The *difference of latitude* (d.lat) between two places is the arc of any meridian contained between the parallels of latitude of the two places. If the two places are both on the north or the south side of the equator the d.lat is found by taking the difference between their latitudes. If the two places are on opposite sides of the equator their d.lat is found by adding their latitudes.

The *difference of longitude* (d.long) between two places is the smaller angle at the pole or the smaller arc of the equator contained between the meridians of the places. If the two places have longitudes of the same name, their d.long is found by subtracting the smaller longitude from the greater. If they lie on opposite sides of the prime meridian their d.long is found by adding their longitudes. If the 180th meridian lies between two places their d.long is found by adding their longitudes and subtracting the sum from 360°.

Although in most cases of navigation it is sufficient to assume the Earth to be perfectly spherical, it is still necessary to consider the true shape of the Earth, principally because the seaman's unit of distance—the *nautical mile*—is related to the Earth's true shape.

The shape of the Earth approximates to that of an *oblate*

spheroid. The surface of an oblate spheroid is traced out by rotating an ellipse about its minor diameter. The Earth's spin axis lies along her least diameter—called the polar diameter. Every meridian is, therefore, an ellipse, and every parallel of latitude is a circle. The diameter of greatest length is the equatorial diameter.

The lengths of the principal radii of the Earth are:

$$\text{Equatorial} = 3963\tfrac{1}{4} \text{ statute miles}$$

$$\text{Polar} = 3949\tfrac{3}{4} \text{ statute miles}$$

The *statute mile* is an arbitrary measure that has evolved from makeshift units such as the length of a barleycorn or the length of a man's foot or span. It is a distance containing 1760 statute yards of defined length. The statute mile is the unit of distance used ashore. For nautical purposes the unit of distance is related to the Earth's dimensions, and is called the nautical mile.

Were the Earth perfectly spherical the length of a minute of arc of her surface would everywhere be the same. This length would provide a useful unit of distance for navigational purposes because it would be equivalent, not only to the length of a minute of longitude at the equator; but, more important, it would be equivalent to the length of a minute of arc of a meridian at any place on Earth: so that, when sailing northwards or southwards, each minute of change of latitude would correspond to a northing or southing of one unit of distance. Although the Earth is not a sphere, this ideal unit of distance does form the basis of measuring distances for navigational purposes.

Now the length of the circumference of any circle of radius R is $2\pi R$. In other words, the radius of a circle fits into the circumference exactly 2π times. The angle at the centre of a circle subtended by an arc of length equal to the radius of the circle is an angular unit known as a *radian* (^{c}), and, clearly:

if $360° = 2\pi^c$

then $1^c = (360 \times 60)/2\pi$ minutes of arc

It follows, therefore, that the radius of a sphere in units of arc length of one minute on its surface is equivalent to the number of minutes of arc in one radian. This is approximately 3438.

Thus, were the Earth a perfect sphere, her radius would be 3438 nautical miles.

The nautical mile is usually defined as a unit of distance equivalent to the length of a minute of arc of a meridian. Because of the elliptical form of the meridians, the nautical mile has a length which varies with latitude. The precise definition of a nautical mile is that it is the length of an arc of a meridian the astronomical latitudes of the end points of which differ by one minute.

The *astronomical latitude* of a place (sometimes called *geographical latitude* or merely *latitude*) is defined as the angle at the place contained between the horizon and the elevated celestial pole measured in the plane of the meridian of the place: the *elevated pole*, as distinct from the *depressed pole*, being the celestial pole lying above the horizon.

Thus two points in the same meridian are one nautical mile apart if the horizons or verticals at the two points are inclined one minute of arc to one another.

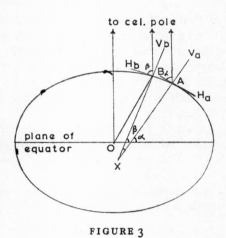

FIGURE 3

Referring to Fig. 3: if V_a and V_b are the verticals and H_a and H_b are the horizontals at two places A and B on the same meridian, the astronomical latitudes of A and B are α and β respectively. If $(\beta - \alpha) = 1'$, then arc AB is one nautical mile.

It follows, therefore, that if the verticals at A and B are inclined to one another one minute of arc, that is, if angle AXB is 1', AB is one nautical mile. It is for this reason that the nautical mile is often described as being the length of a minute of arc of a meridian subtended by an angle of one minute at the centre of curvature of the meridian.

The radius of curvature of any part of a meridian varies with latitude, being least at the equator and greatest at the pole. This results in the length of the nautical mile being greatest at the poles and least at the equator.

Numerous attempts have been made to ascertain the exact size and shape of the Earth. This is expressed in terms of the ellipticity or compression and the equatorial radius, forming the so-called *figure of the Earth*. The *ellipticity*, *c*, of the Earth is the ratio between the difference of the lengths of the equatorial and polar radii, and the length of the equatorial radius. Thus if the equatorial and polar radii are denoted by *a* and *b* respectively:

$$c = (a - b)/a$$

The figure of the Earth used as a basis for Admiralty charts is that of the Clarke spheroid of 1880, which has an ellipticity of 1/293·5 and an equatorial radius of 20,925,972 feet. On this spheroid the length of a nautical mile in latitude 0° is 6046 feet, and at latitude 90° it is 6108 feet.

The average length of a nautical mile on the Clarke (1880) spheroid is 6077 feet.

The standard nautical mile adopted in Great Britain is 6080 feet or 1852 metres, which is the length of the actual nautical mile in latitude 48°. This is the nearest round figure to the length of the average nautical mile.

The general expression for finding the length of the nautical mile in any latitude ϕ is:

$$(6077 - 31 \cos 2\phi) \text{ feet}$$

The standard nautical mile is used as a basis for regulating patent logs. No material error is introduced into normal navigational problems by using the standard nautical mile for all latitudes. The greatest error occurs in very low latitudes; but, even here, the error amounts to no more than about 1 mile in 50.

Since error in the estimated effects of wind and current normally exceed this, the discrepancy due to the standard nautical mile being different from that of the actual nautical mile escapes notice in practical navigation.

Assuming the earth to be perfectly spherical, it is an easy matter to prove that the length of any parallel of latitude is proportional to the cosine of the latitude. If we regard a mile as the unit length which is equivalent to an arc length of one minute on the surface of a spherical Earth, the distance in miles between any two points on the same parallel of latitude is equal to the product of the d.long between the points and the cosine of the latitude. That is:

$$d = D \cos \phi$$

where d = distance in miles between two points on same parallel

D = d.long in minutes of arc between the two points

ϕ = latitude of points.

This is the *Parallel Sailing Formula*, used extensively in navigation.

The distance in miles that one point is east or west of another is known as the *departure* between the two points.

CHAPTER III

The Celestial Sphere

The celestial sphere is that sphere on to which the heavenly bodies are assumed to be projected in order to facilitate the study of the apparent motions of the heavenly bodies. Because the celestial sphere has infinite radius it matters not whether we assume the point of projection to be the observer's eye, the Earth's centre, or even, for some purposes, the Sun.

The stars, because of their immense distances from the Solar System, tend to maintain their positions on the celestial sphere; but the Sun and other members of the Solar System, because of their relative nearness to the Earth, and because the Earth rotates and revolves, change their positions on the celestial sphere relative to the stars comparatively rapidly. Let us consider the apparent annual movement of the Sun on the celestial sphere due to the Earth's orbital motion.

The period of the Earth's revolution around the Sun is a year; so that in one year the Sun appears to describe a great circle on the celestial sphere which is co-planar with the Earth's orbit. The great circle forming the Sun's annual apparent path is called the *ecliptic*. In Fig. 1, which illustrates the celestial sphere, the Earth's orbit, and the ecliptic, the Sun is assumed to occupy the central position of the celestial sphere.

When the Earth is at points a, b and c, shown in Fig. 1, the Sun is projected on to the celestial sphere at positions A, B and C respectively. Thus, as the Earth moves in her orbit from a to c, the Sun appears to move across the celestial sphere from A to C.

The plane of the Earth's rotation is inclined to the plane of her orbit at an angle of about $23\frac{1}{2}°$ so that the *equinoctial*, which is the projection of the equator on to the celestial sphere, is inclined at an angle of $23\frac{1}{2}°$ to the plane of the ecliptic. This angle is called the *Obliquity of the ecliptic.*

The Earth's spin axis is inclined at an angle of $(90 - 23\frac{1}{2})°$, that is $66\frac{1}{2}°$, to the plane of her orbit around the Sun; and, because of the Earth's gyroscopic inertia, this angle tends to be constant. It follows, therefore, that for half the year the Earth's

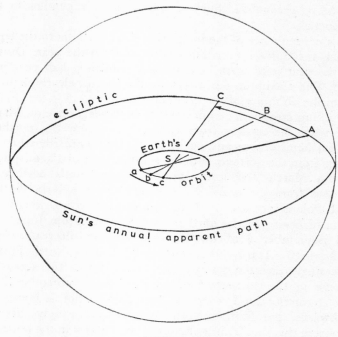

FIGURE I

North Pole is directed towards the Sun; and for the remaining half year it is directed away from the Sun. Also, during half the year the Sun lies in the northern celestial hemisphere, and for the remaining half he lies in the southern celestial hemisphere.

A heavenly body in the northern celestial hemisphere is said to have *north declination*, and one in the southern celestial hemisphere *south declination*. The declination of a heavenly body is its angular distance north or south of the equinoctial and is analogous to latitude on the terrestrial sphere—this being angular distance north or south of the equator. Semi-great circles extending from the north to the south celestial pole cross the equinoctial at right angles—these are called *celestial meridians*.

All points on the celestial sphere having the same declination (either north or south) lie on a small circle which is called a *parallel of declination*. Such a small circle is parallel to the equinoctial.

As a consequence of the inclination of the ecliptic to the equinoctial, the Sun's declination changes during the year. During half the year it is north, and for the remaining half year it is south. The maximum declination of the Sun is, clearly, equivalent to the obliquity of the ecliptic.

The Sun crosses the equinoctial twice a year: once on March 21st, when he crosses from the southern into the northern celestial hemisphere; and again on September 23rd, when he crosses from the northern into the southern celestial hemisphere.

From March 21st until June 22nd the Sun's declination increases from 0° to 23½°N. During this period the Earth's North Pole is increasingly directed towards the Sun: this is the season of *Spring* in the northern hemisphere. From June 22nd until September 23rd the Sun's declination decreases from 23½°N. to 0°. During this period the Earth's North Pole is decreasingly directed towards the Sun. This is the season of *Summer* in the northern hemisphere. From September 23rd until December 22nd, when the Sun's declination is increasing southwards, the Earth's North Pole is increasingly directed away from the Sun. This is the season of *Autumn* in the northern hemisphere. The remaining quarter of the year, from December 22nd to March 21st, when the Sun's declination decreases from 23½°S. to 0°, the Earth's North Pole is decreasingly directed away from the Sun. This is the season of *Winter* in the northern hemisphere. In the southern hemisphere the seasons are the reverse from what they are in the northern hemisphere.

Fig. 2 illustrates the seasons.

The great circle on the Earth which divides the dark from the Sun-enlightened hemisphere is called the *circle of illumination*. The Earth's axis lies in the plane of the circle of illumination on March 21st and September 23rd. On these days, therefore, day and night are each 12 hours all over the Earth. For this reason these days are called the *Spring* or *Vernal*, and the *Autumnal equinoxes* respectively. The points on the celestial sphere occupied by the Sun when his declination is 0° are called the *equinoctial points*.

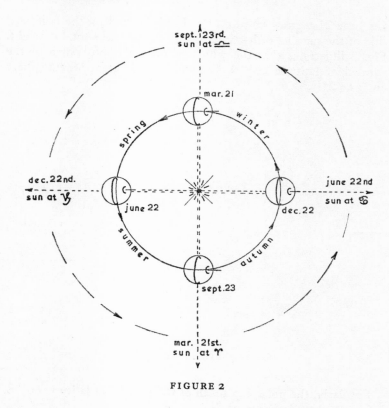

FIGURE 2

On June 22nd and December 22nd the Earth's spin axis is inclined at a maximum angle of 23½° to the plane of the circle of illumination, and the length of daylight is maximum for the northern and southern hemispheres respectively. On these days, the Sun's declination ceases to increase and he appears to stand still relative to the equinoctial. For this reason the times when the Sun reaches his maximum declination are called *solstices*, and the points on the celestial sphere occupied by the Sun at these times are called *solstitial points*.

UNEQUAL LENGTHS OF DAYLIGHT AND DARKNESS DURING THE YEAR

On June 21st every point on the Earth located on the poleward side of the parallel of latitude $66\frac{1}{2}$°N. experiences total daylight. Fig. 3 illustrates that the whole of the polar cap north of the parallel of latitude $66\frac{1}{2}$°N. lies in the enlightened hemisphere on June 21st.

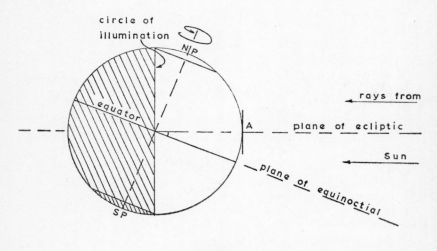

FIGURE 3

Similarly, the polar cap south of the parallel of latitude $66\frac{1}{2}$°S. is in complete darkness on June 21st.

It will be noticed from Fig. 3 that at noon on June 21st the Sun is vertically overhead to an observer at A located on the parallel of latitude $23\frac{1}{2}$°N. At all places north of this parallel, the Sun crosses the plane of an observer's meridian bearing south; and at all places to the south of this parallel the Sun crosses the plane of an observer's meridian bearing north.

An observer at any place to the north of the equator on the day of the Summer solstice, is on the enlightened side of the circle of illumination for a period of longer than 12 hours. An observer at any place to the south of the equator is on the dark side of the circle of illumination for longer than 12 hours on this day.

It may be readily verified that on the day of the Winter solstice (December 22nd) all places north of the parallel of latitude $66\frac{1}{2}°$N. experience total darkness, and all places south of the parallel of latitude $66\frac{1}{2}°$S. experience total daylight. All places in the southern hemisphere have daylight for more than 12 hours each day, and all places in the northern hemisphere have daylight for less than 12 hours.

On the days of the equinoxes, day and night are each 12 hours at all places on Earth, and the Sun crosses the plane of the meridian with an altitude of 90° at every point on the equator.

CLIMATIC ZONES

As a result of the changing declination of the Sun the Earth's surface is divided into climatic zones bounded by the parallels of latitude 0°, $23\frac{1}{2}°$ and $66\frac{1}{2}°$ N. and S.

The zone that lies between the equator and the parallel of $23\frac{1}{2}°$N. is called the *North Torrid Zone*. That which lies between the equator and $23\frac{1}{2}°$S. is called the *South Torrid Zone*. Within the Torrid zone, which is bounded by the *Tropic of Cancer* in the north and the *Tropic of Capricorn* in the south, the Sun has an altitude of 90° on two days every year.

The parallel of latitude $66\frac{1}{2}°$N. is called the *Arctic Circle*, and that of $66\frac{1}{2}°$S. is called the *Antarctic Circle*. Within the polar caps north of the Arctic Circle and south of the Antarctic Circle, at least one day of the year has 24 hours daylight and at least one day of the year has 24 hours complete darkness.

The zones lying between the Tropic of Cancer and the Arctic Circle, and the Tropic of Capricorn and the Antarctic Circle, are called the *North* and *South Temperate Zones* respectively. The Earth's climatic zones are illustrated in Fig. 4.

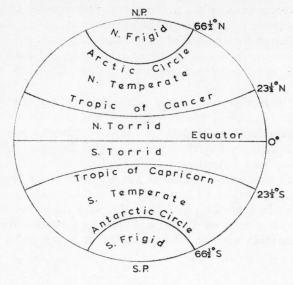

FIGURE 4

UNEQUAL LENGTHS OF THE SEASONS

The orbital speed of the Earth varies, being greatest when she is nearest to the Sun and least when she is farthest from him. The Earth is at perihelion on January 3rd approximately. This is a fortnight or so after the Winter solstice. From the time of the Winter solstice to that of the Spring equinox the Earth travels faster than at the other seasons. The effect of this is to bring forward the first day of Spring in the northern hemisphere.

The Summer solstice occurs about a fortnight before the time of aphelion (about July 3rd). The Earth, during northern Summer, therefore, moves slower than during the other seasons. The effect of this is to delay the first day of Northern Autumn. The seasons, therefore, do not have equal lengths. For the northern hemisphere:

Spring	is	93	days (approx.)
Summer	is	94	,,
Autumn	is	90	,,
Winter	is	89	,,

Because northern Winter occurs when the Earth is comparatively near the Sun, the severity of northern Winter is mitigated. Similarly, southern Winter, which takes place when the Earth is comparatively far from the Sun, is correspondingly harsher than it would be were the seasons of equal length.

THE ZODIACAL BELT

During a year the Sun travels through twelve constellations which lie within a celestial zone centred along the ecliptic. This zone is called the *zodiacal belt*, and the twelve constellations, the *Signs of the Zodiac*.

At the time when the Sun's apparent annual path was investigated by the Egyptian astronomers of antiquity, the Sun entered the constellation of Aries the Ram on the first day of Spring. For this reason the Spring equinoctial point is called the *First Point of Aries* (denoted by the symbol ♈). A month later the Sun entered the next sign of the zodiac, and was said to be at the first point of Taurus the Bull, etc. The twelve signs of the zodiac are usually memorized from the following rhyme:

> The Ram, the Bull, the Heavenly Twins,
> And next the Crab the Lion shines,
> The Virgin and the Scales,
> The Scorpion, Archer and He-goat,
> The Man who holds the watering pot,
> And the Fish with the glittering tails.

The signs of the zodiac are illustrated in Fig. 5.

Because of the precession of the Earth's polar axis the equinoctial points are no longer at the first points of Aries and Libra, although the names are still used.

Precession of the axis of a spinning body takes place when an external couple acts upon it. When a spinning body precesses, its axis traces out a conical surface taking a time which is usually very slow compared with its period of rotation.

The precession of the Earth's axis is due to:

1. The Earth's oblate shape.
2. The tilt of the spin axis to the plane of the ecliptic.

3. A torque acting due to solar attraction which tends to force the Earth's plane of rotation into that of its revolution around the Sun.

Precession of the Earth's axis results in the precession of the celestial poles and the equinoctial points. The celestial poles

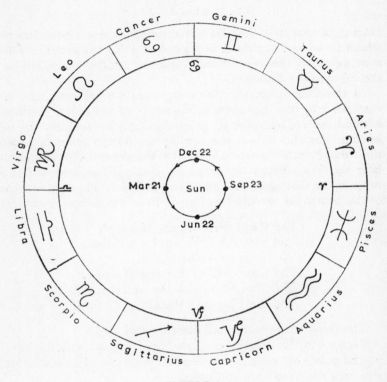

FIGURE 5

describe small circles centred at the poles of the ecliptic. The period of the precession of the celestial poles is 26,000 years. This slow motion results in the equinoctial points moving with retrograde motion around the ecliptic at an angular rate of about 50 seconds of arc per year. This movement is called the *precession of the equinoxes.*

The Moon, the plane of whose orbit is inclined at an angle of about $5\frac{1}{4}°$ to the plane of the Earth's orbit, has a similar effect on the Earth as has the Sun. The points of intersection of the planes of the Moon's and Earth's orbits are known as the *nodes*. The point on the celestial sphere which the Moon occupies when she crosses from the south to the north of the ecliptic is called the *ascending node*. The other is called the *descending node*. The straight line joining the nodes, known as the *nodal line*, swings westwards around the ecliptic with a period of about 19 years.

When the ascending node coincides with the Spring equinox the maximum declination of the Moon is $(23\frac{1}{2} + 5\frac{1}{4})°$, that is $28\frac{3}{4}°$ N. or S. About $9\frac{1}{2}$ years later, when the descending node coincides with the Spring equinox, the maximum declination of the Moon is $(23\frac{1}{2} - 5\frac{1}{4})°$, that is $18\frac{1}{4}°$ N. and S. When the nodes coincide with the solstitial points the declinational limits of the Moon are the same as those of the Sun.

The Earth's axis does not describe a smooth conical surface, as it would were precession due to the Sun alone. The Moon's effect, known as *nutation*, is to cause the celestial pole to trace out a wavy curve, each wave taking about 19 years to complete.

THE MOON

The Moon's orbit around the Earth is elliptical. When she is at perigee the Earth–Moon distance is some 220,000 miles; and when she is at apogee it is about 253,000 miles.

As the Moon revolves in her orbit around the Earth she describes a great circle on the celestial sphere at an angular rate of about 13° per day. This comparatively rapid apparent motion may readily be observed over a short period of time. To complete a circuit around the celestial sphere relative to the stars the Moon takes about $\frac{360}{13}$, that is about $27\frac{1}{3}$ days. This period is known as a *sidereal period*.

The Moon is rendered visible by reflected sunlight. The proportion of the Moon's illuminated hemisphere visible at the Earth at any time is dependent upon the relative positions of the Sun, Moon and Earth. The changing shapes of the Moon's surface observed at the Earth are known as the Moon's *phases*.

When the Moon is in conjunction, her illuminated hemisphere is directed away from the Earth. At this time the Sun and Moon cross the plane of the meridian of an observer together at noon. At this time the Moon is said to be *New*, and her *age* is said to be 00 days 00 hours.

As the angle at the Earth between the Moon and the Sun increases after the time of New Moon, the phase changes from a crescent form to *Half Moon* which occurs about seven days after New Moon. At this time the Moon is said to be at *First Quarter*. When the shape is greater than Half Moon the phase is said to be *gibbous*. When the age of the Moon is 14 or 15 days, at the time when she is in opposition with the Sun, the full illuminated disc is visible at the Earth. The Moon is then said to be *Full*.

From conjunction to opposition with the Sun the phase of the Moon changes from New to Full. During this time the Moon is said to *wax* and her western limb is illuminated. After Full Moon the illuminated part of the Moon's surface visible at the Earth diminishes and the Moon is said to *wane* during which time her eastern limb is illuminated. Seven days after Full Moon, when half the illuminated hemisphere is visible at the Earth, the Moon is said to be at the *Third Quarter*. After a further seven days, the Moon is said to be at *Change*, when her age is again 00 days 00 hours.

The period of the phases of the Moon, that is to say the interval between successive New Moons, is about $29\frac{1}{2}$ days. This is a couple of days longer than a sidereal period, because the daily separation of the Sun and Moon is about 12°, that is 1° less than the arc traced out by the Moon relative to the fixed stars. The *lunation* is, therefore, $\frac{360}{12}$ or $29\frac{1}{2}$ days. The lunation is sometimes called a *synodic period*.

Twelve lunations amount to 354 days, which is about 11 days less than a solar year, so that the age of the Moon on January 1st of successive years increases by 11. The age of the Moon on January 1st is called the *epact* for that year. This is used for ascertaining the date of Easter in the ecclesiastical calendar.

WINTER AND SUMMER FULL MOONS

Successive Full Moons take place in different parts of the celestial sphere: this is due to the Sun's annual apparent motion across the celestial sphere.

In Summer, when the Sun's declination is north, the Full Moon has south declination. In the northern hemisphere, because celestial bodies having south declination are above the horizon for less than 12 hours each day (see Chapter V), Summer Full Moons are above the horizons of northern hemisphere observers for less than half a lunar day. In contrast, Winter Full Moons, having north declination, are above the horizons of northern hemisphere observers for more than half the lunar day.

In Spring and Autumn, when the Full Moon occurs near the First Points of Aries and Libra respectively, the Moon's declination changes most rapidly during the lunation.

In Spring, when the Sun's declination changes from south to north, the Full Moon's declination changes from north to south. This has the effect of accelerating the time of Moonset. In Autumn, in contrast, and for similar reasons, the time of Moonset is retarded. The Full moon immediately following the Autumnal equinox is called the *Harvest Moon*. The interval between the times of Sunset and Moonrise is short for several days after Harvest Full Moon on account of the rapidly changing declination of the Moon and its effect on delaying the time of Moonrise; so that, before darkness sets in, the large Moon rises to provide reflected sunlight for the harvest gatherers.

MOON'S LIBRATIONS

The Moon rotates once in a sidereal period, the speed of rotation being uniform. Because of this, the same side of the Moon is always directed towards the Earth.

The Moon's orbital motion is not uniform, her speed of revolution around the Earth being greatest at perigee and least at apogee. The average orbital speed is equivalent to the rotational speed, so that as perigee is approached a narrow lune on the western side of the circle of illumination on the Moon becomes visible; and a narrow lune on the eastern side becomes invisible, swinging, as it does, into the dark side of the circle of illumination. Similarly, when the Moon is approaching apogee, the rotational speed being greater than the orbital speed, a narrow lune, normally beyond the eastern limb, heaves into view.

The Moon's axis of rotation is inclined at an angle of 84° to the plane of its orbit around the Earth. During a sidereal period, therefore, terrestrial observers are able to see a 6° lune extending over and under the normal polar limbs.

The apparent shaking and nodding of the Moon, due to the above factors, are known as *librations in longitude* and *latitude* respectively. They result in it being possible to observe from the Earth about 60 per cent of the Moon's surface instead of a little less than half, which would be the case if the Moon had no librations.

ECLIPSES

Should the celestial positions of the Moon and Sun be the same, the Sun's disc would be obscured by the Moon. This phenomenon is called a *solar eclipse*.

Should the celestial positions of the Moon and Sun be diametrically opposed on the celestial sphere, the Moon would be obscured by the shadow of the Earth cast by the Sun. This phenomenon is called a *lunar eclipse*.

An eclipse of the Sun occurs at the time of New Moon, and an eclipse of the Moon occurs when the Moon is Full.

Were the Earth's orbit around the Sun and the Moon's orbit around the Earth co-planar, an eclipse of the Sun and an eclipse of the Moon would occur once during every lunation. Because the plane of the Moon's orbit is inclined at an angle of $5\frac{1}{4}°$ to the plane of the Earth's orbit, eclipses do not occur so frequently as they would were the two orbits co-planar. For an eclipse to occur the Moon must lie on or near to the ecliptic; and this is the reason why the projection of the Earth's orbit on the celestial sphere is called the ecliptic.

Eclipse information is tabulated in the seaman's *Nautical Almanac*. For 1968, for example, we learn from the *Nautical Almanac* that there were four eclipses, two of the Sun and two of the Moon. Diagrams illustrating the times and limits within which solar eclipses are visible are also provided in the *Nautical Almanac*.

Lunar eclipses may be *partial* or *total*, according to whether the Full Moon is partially or completely obscured in the Earth's shadow.

Solar eclipses may be *partial*, *total* or *annular*. A total or annular eclipse of the Sun is visible only within a relatively narrow strip of the Earth's surface, the width of the strip being never more than about 170 miles. This narrow strip is called the *path of the eclipse*. Within wider strips of territory, up to about 1000 miles, adjacent to the path of an eclipse, a partial eclipse of the Sun may be observed. Within the path of the eclipse the Sun is completely obscured to form a total eclipse if the angular diameter of the Moon exceeds that of the Sun. If the Sun's angular diameter is greater than that of the Moon's, within the path of an eclipse a narrow circumferential area of the Sun's disc will be visible, in which case the eclipse is described as annular.

The Moon, during her monthly circuit of the heavens, frequently passes over fixed stars and occasionally over planets which lie in her path. When this happens the star or planet is said to be *occulted*, and the phenomenon is called an *occultation*.

CHAPTER IV

On Defining Celestial Positions

The most satisfactory method of defining a position on a plane surface is by employing Cartesian co-ordinates relative to mutually perpendicular axes of references in the plane. The same principle is used for defining positions on a sphere: the axes of reference, in this case, are two great circles which intersect at right angles.

There are three systems of defining celestial positions. These are:

1. The Ecliptic system.
2. The Equinoctial system.
3. The Horizon system.

THE ECLIPTIC SYSTEM

The names given to the co-ordinates used in the ecliptic system of defining celestial positions are *celestial latitude* and *celestial longitude* respectively. The two great circles of reference from which celestial longitudes and celestial latitudes are measured are the ecliptic, which gives its name to the system, and a secondary great circle to the ecliptic which extends from the extremities of the axis of the ecliptic and which passes through that intersection of the ecliptic and equinoctial called the First Point of Aries. The extremities of the axis of the ecliptic are called the *poles of the ecliptic* and the semi-great circles extending between them are called *circles of latitude*.

The celestial latitude of a celestial point is a measure of the

44

arc of a circle of latitude between the point and the ecliptic. The celestial latitude of a point is named north or south according to whether the point lies north or south of the ecliptic respectively. The celestial latitude of any point on the ecliptic is 0°. The celestial latitude of the Sun, therefore, is always 0°.

The celestial longitude of a celestial point is a measure of the arc of the ecliptic contained between the First Point of Aries and the circle of latitude through the point, measured eastwards from 0° to 360° from the First Point of Aries. The celestial longitude of the Sun is 0° at the time of the Spring equinox, and it increases at an irregular rate until the following Spring equinox.

Because of the precession of the equinoxes the celestial longitudes of all fixed points on the celestial sphere increase with time. The celestial latitudes of all fixed points on the celestial sphere are constant.

The ecliptic system of defining celestial positions is useful in a consideration of the *Equation of Time* (see Chapter VI).

THE EQUINOCTIAL SYSTEM

The co-ordinates used in the equinoctial system of defining celestial positions are *declination* and *Right Ascension* (R.A.), or some other similar angle. The two great circles of reference from which declination and R.A. are measured are the equinoctial and the secondary to the primary equinoctial which connects the extremities of the axis of the equinoctial and which passes through the First Point of Aries. The extremities of the axis of the equinoctial are the *celestial poles*, and semi-great circles extending between them are called *celestial meridians*.

The declination of a celestial point is a measure of the arc of a celestial meridian contained between the point and the equinoctial. The equinoctial divides the celestial sphere into the northern and southern hemispheres. All points in the northern celestial hemisphere have north declination; and all points in the southern celestial hemisphere have south declination. All points having the same declination lie on a small circle which is parallel to the equinoctial. Such a small circle is a *parallel of declination*. The declination of any point on the equinoctial is 0°; and the declination of the celestial pole is 90°. The Sun's declination changes from 0° to 23½° N., to 0° to 23½° S., to 0° again in the course of a year.

The angle between a celestial point and the elevated celestial pole is equivalent to the complement of the declination of the point if both point and celestial pole lie in the same celestial hemisphere. If they lie in opposite hemispheres the angle between the point and the pole is equivalent to (90° + declination). The distance between a point on the celestial sphere and the elevated celestial pole is called the *polar distance* of the point.

The Right Ascension of a celestial point is a measure of the arc of the equinoctial, or the angle at the celestial pole, contained between the celestial meridians of the point and the First Point of Aries. It is always measured in hours, minutes and seconds, from the celestial meridian of the First Point of Aries.

The R.A. of the Sun is 00 hr 00 min 00 sec when he is at the First Point of Aries at the time of the Spring equinox. The Sun's R.A. increases irregularly from 00 hr 00 min 00 sec on March 21st to 24 hr 00 min 00 sec on the following March 21st.

Because of the precession of the equinoxes the declination and R.A. of every fixed point on the celestial sphere change with time.

Fig. 1 illustrates the ecliptic and equinoctial systems of defining celestial positions.

In Fig. 1, the circle represents the celestial sphere. ♈ represents the First Point of Aries and X is any celestial body.

Using the ecliptic system:

$$X \text{ is in } \begin{cases} \text{celestial latitude} &= \text{arc BX} \\ \text{celestial longitude} &= \text{arc ♈B} \end{cases}$$

Using the equinoctial system:

$$X \text{ is in } \begin{cases} \text{declination} &= \text{arc AX} \\ \text{R.A.} &= \text{arc ♈A} \end{cases}$$

For navigational purposes the celestial positions of the navigational stars are given in the *Nautical Almanac* in terms of declination and *Sidereal Hour Angle* (S.H.A.).

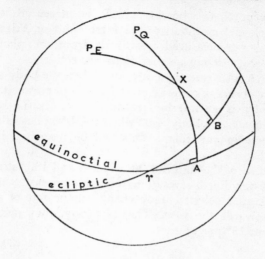

FIGURE I

The sidereal hour angle of a star is a measure of the arc of the equinoctial or the angle at the celestial pole, contained between the celestial meridians of the First point of Aries and the star. In contrast to R.A., S.H.A. is measured westwards from the celestial meridian of the First Point of Aries, so that:

$$S.H.A. * = 360° - R.A. * \text{ in degrees.}$$

Time problems in nautical astronomy are facilitated by using S.H.A. instead of R.A., principally because the S.H.A. of a celestial body is measured in the same direction as the apparent diurnal motion of the celestial sphere, a motion which will be discussed in detail in Chapter V.

Declination and R.A. are co-ordinates used for defining celestial positions relative to fixed great circles in the celestial sphere. It is convenient to provide a method of defining celestial positions relative to the projection of the observer's (and the prime) meridian on to the celestial sphere. For this purpose use is made of co-ordinates called *Local* and *Greenwich Hour Angle* (L.H.A. and G.H.A.).

The L.H.A. or G.H.A. of a celestial position is similar to

S.H.A. except that, whereas S.H.A. is measured westwards from the celestial meridian of the First Point of Aries, L.H.A. and G.H.A. are measured westwards from the plane of the Observer's and Greenwich meridians respectively.

The L.H.A. of a celestial body is 0° when the body lies in the plane of an observer's meridian at its superior transit. When a celestial body is at inferior transit its L.H.A. is 180°. The G.H.A. of a celestial body is 0° when the body lies in the plane of the Greenwich meridian. This will be discussed in greater detail in Chapter VI.

The L.H.A. of the Sun is 0° at midday, and 180° at midnight. Because of the Earth's rotation towards the east, the L.H.A. and G.H.A. of a celestial body change at the rate of about 15° per hour towards the west. That is to say, they increase at a rate of about 15° per hour.

THE HORIZON SYSTEM

The co-ordinates used in the Horizon system of defining celestial positions are *altitude* and *azimuth*. The two great circles of reference from which these angles are measured are the *celestial horizon* and a secondary to the horizon which extends between the poles of the primary horizon and which crosses it at the north and south points of the horizon.

The extremities of the axis of the horizon are called the *zenith* and *nadir* respectively. The celestial horizon divides the celestial sphere into the *visible* and *invisible hemispheres*. The zenith is the pole of the horizon which lies in the visible hemisphere. The nadir is antipodal to the zenith. Secondary great circles to the horizon are called *vertical circles*.

The altitude of a celestial point is a measure of the arc of a vertical circle between the point and the horizon vertically below it. The altitude of every point on the horizon is 0°, and the altitude of the zenith is 90°. All points having the same altitude lie on a small circle which is parallel to the horizon. Such a small circle is called a *parallel of altitude*.

The angle contained between a celestial point and the zenith is called the *zenith distance* (Z.D.) of the point. If the point is above the horizon, the zenith distance is equivalent to the complement of the altitude of the point. If it lies below the horizon, its zenith distance is greater than 90°.

The vertical circle that extends from the zenith of an observer to the north point of his horizon lies in the plane of the observer's meridian. It, therefore, contains the celestial pole of the hemisphere in which the observer is located. That is to say, if the observer is in the northern hemisphere, the north celestial pole will lie on the vertical circle through the north point of the horizon, and it will bear due north. If the observer is in the southern hemisphere the south celestial pole will lie above the horizon bearing due south.

The celestial pole lying above an observer's horizon is called the *elevated pole*; the other is called the *depressed pole*.

Semi-great circles on the celestial sphere which terminate at the celestial poles are *celestial meridians*. It follows, therefore, that the vertical circle through the elevated pole is also a celestial meridian; and, containing as it does, the observer's zenith, it lies in the plane of the observer's terrestrial meridian. For this reason it is called the *observer's celestial meridian*.

The *azimuth* of a celestial point is a measure of the arc of the horizon, or the angle at the observer's zenith, contained between the vertical circle through the elevated pole and the vertical circle through the point. It is named east or west according to whether the point lies to the east or west of the observer's celestial meridian. Azimuths of celestial points are measured from north or south according to whether the observer is in the northern or southern hemisphere respectively. The azimuth of every point, the declination of which is named opposite to that of the latitude, is greater than 90°. It does not follow, however, that objects having declinations of the same name as that of the latitude have azimuths of less than 90°. For the azimuth of an object to be less than 90° its declination must be greater than, but of the same name as, the latitude.

The vertical circle passing through the east and west points of the horizon is called the *prime vertical circle*. The azimuth of every point on the prime vertical circle is N. or S. 90° E. or W. according to whether it is east or west of the observer's celestial meridian.

All objects having azimuths named east are *rising* objects;

that is, their altitudes are increasing. All objects on the west side of the observer's celestial meridian are *setting* objects; that is, their altitudes are decreasing. When a fixed celestial object crosses a stationary observer's celestial meridian it has its greatest altitude for the day (see Chapter V).

The *bearing* of a celestial object, as distinct from its azimuth, is a measure of the horizontal angular distance from the vertical plane through the north or south point of the horizon. For convenience it is named from the nearer cardinal point. For example, if the azimuth of a body is say N. 120° E., its bearing would be described as S. 60° E.

The *amplitude* of a heavenly body is a measure of its angular distance from the east or west point of the horizon and the body when it rises or sets respectively. Rising amplitudes are named from the east point of the horizon; and setting amplitudes are named from the west point. Celestial bodies having north declination rise and set north of the prime vertical circle; and those having south declination rise and set south of the prime vertical circle. In other words the amplitude of a body takes its name from the declination of the body.

Fig. 2 illustrates the horizon system of defining celestial positions. Fig. 2(a) illustrates the celestial sphere drawn on the plane of the horizon of an observer whose zenith is at Z. N, E,

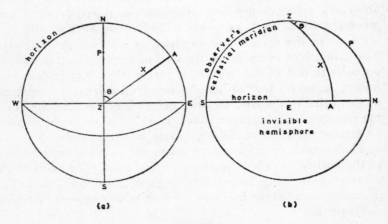

(a)   (b)

FIGURE 2

S and W are the cardinal points of the observer's horizon. P is the elevated celestial pole, and X is a celestial body.

Fig. 2(b) illustrates the celestial sphere drawn on the plane of the observer's celestial meridian. Arc PZS is part of the observer's upper celestial meridian, arc PN is part of his inferior celestial meridian.

In both diagrams:

Position of X using the Horizon system:

Altitude = arc AX
Azimuth = NZX or arc NA
(Bearing = N. $\theta°$ E.)

CHAPTER V

The Apparent Diurnal Motion of the Celestial Sphere

The Earth's diurnal rotation is manifested by the apparent diurnal revolution of the celestial sphere. As a result of this during the course of a day all celestial objects rise out of the eastern half of an observer's horizon and set into the western half each day.

To terrestrial observers, the celestial objects appear to describe paths around the Earth, each sweeping out a complete circle once per day. These circles are called *diurnal circles*. Let us first consider the diurnal circles of the heavenly bodies as viewed by each of two observers, one located at the North Pole of the Earth, and the other located at the equator. We shall then discuss diurnal circles as viewed by an observer located in some intermediate latitude.

Fig. 1 illustrates the celestial sphere with the Earth lying at its centre. The zenith Z, of an observer standing at the Earth's North Pole, coincides with the north celestial pole P. Now the celestial pole is the pole of the equinoctial, and the zenith is the pole of the horizon, so that if the pole and the zenith coincide, so also will the equinoctial and the horizon.

Now the diurnal revolution of the celestial sphere takes place about the axis of the equinoctial. Therefore, to an observer at the Earth's North Pole every point on the celestial sphere will trace out a diurnal circle which is parallel to his horizon. It follows, therefore, that every fixed celestial object will maintain a constant altitude to an observer located at the Earth's North Pole, and will trace out its diurnal circle in an anti-clockwise direction as viewed by the observer.

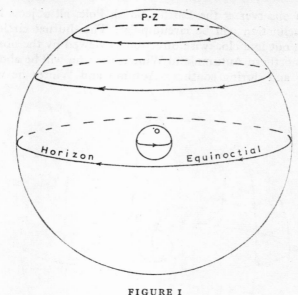

FIGURE I

The equinoctial, which coincides with the horizon of an observer located in latitude 90°, divides the celestial sphere into hemispheres of north and south declination. At the Earth's North Pole all celestial points having north declination lie above the horizon. Those having south declination lie below the horizon in the invisible hemisphere.

Celestial bodies which are above the horizon throughout the day are called *circumpolar bodies*. To an observer at the North Pole of the Earth all bodies having north declination are circumpolar. The Sun will be circumpolar during the period from March 21st to September 23rd. During the three months period between March 21st and June 22nd, that is, during the season of northern Spring, the Sun's altitude will change gradually from 0° to $23\frac{1}{2}$°, attaining its maximum altitude on the day of the Summer solstice. After this day for the following three months, that is, during the season of northern Summer, the Sun's altitude will decrease to 0°. During the remaining half year, the Sun will never rise above the horizon of an observer at the Earth's North Pole.

To an observer at the Earth's South Pole, all objects having south declination will be circumpolar. The diurnal circles will be swept out in a clockwise direction as viewed by the observer. During northern Autumn and Winter, the Sun will be above the horizon; and during southern Autumn and Winter he will be below it.

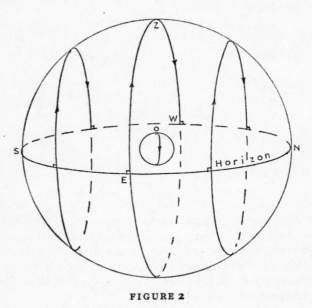

FIGURE 2

Fig. 2 illustrates the celestial sphere with the Earth at the centre. o is an observer located on the equator. Z is his zenith, and N, E, S and W are the cardinal points of the observer's horizon.

It will be seen from Fig. 2 that the horizon of an observer located on the equator bisects the equinoctial. It follows, therefore, that the diurnal circles of all celestial bodies are bisected by the horizon. All celestial bodies, therefore, rise and set to an observer located on the equator; and no bodies are circumpolar.

It will also be noticed from Fig. 2 that all diurnal circles cross the equinoctial perpendicularly. Moreover, all objects having north declination will rise to the north of east and set to the north of west, and their azimuths at all times will be less than 90° measured from the north point of the horizon. Similarly, all objects having south declination will rise to the south of east and set to the south of west, and their azimuths also will be less than 90° measured from the south point of the horizon. A celestial body having a declination of 0° will rise bearing due east and set bearing due west. Such an object will cross the plane of the observer's meridian with an altitude of 90°.

When a celestial object crosses the observer's celestial meridian it is said to *culminate*. To an observer at the equator, every celestial body will culminate with an altitude equal to the complement of its declination. Because every diurnal circle is bisected by the horizon of an observer located on the equator, every celestial body will be above the horizon for exactly half the day. Moreover it will culminate six hours after rising and set six hours after culminating.

Fig. 3 illustrates the celestial sphere with the Earth at the

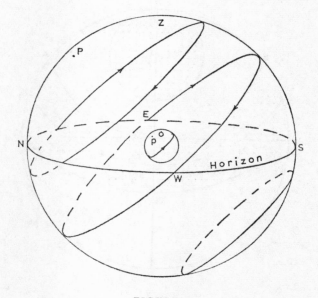

FIGURE 3

centre. o is an observer in the northern hemisphere. Z is the observer's zenith and N, E, S and W are the cardinal points of his horizon. p is the Earth's North Pole and P is the celestial north pole.

In this general case it will be noticed that the planes of all diurnal circles lie at an angle to that of the horizon. It will also be noticed from Fig. 3 that some objects having north declination are above the horizon all day, and that some having south declination are below the horizon all day. The number of circumpolar bodies depends upon the observer's latitude: the higher the latitude the greater the number. Celestial bodies having north declination will be above the horizon of an observer having north latitude for longer than 12 hours each day. Those having south declination will be above the observer's horizon for less than 12 hours each day.

Fig. 4 illustrates the celestial sphere. The boundary circle represents the horizon, and the points on the celestial sphere

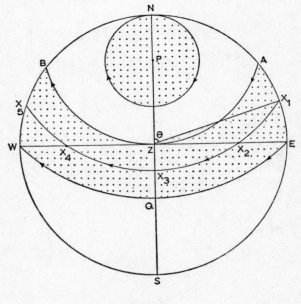

FIGURE 4

in which we are interested are projected on to the plane of the horizon. Z represents the observer's zenith. PZS is the projection of the observer's upper celestial meridian, and PN is that of his lower celestial meridian. EQW is the projection of the equinoctial—every point on which is 90° from P the projection of the celestial pole.

All diurnal circles are parallel to the equinoctial; and those within the shaded circle centred at P are circumpolar. It will be noticed from Fig. 4 that all objects within the shaded zone bounded by the diurnal circle AZB and the equinoctial will, at two instants in the day, cross the observer's prime vertical circle.

Consider the star X which rises at X_1. Its rising amplitude is arc EX_1: its azimuth being N. θ E. where θ is angle PZX_1. The diurnal circle of X crosses the prime vertical circle at X_2 and X_4, at which times its azimuth is N. 90° E. and N. 90° W. respectively. When it culminates it is at X_3 and bears due south with its greatest altitude for the day being equivalent to arc SX_3. At X_5 it sets with a setting amplitude of arc WX_5.

The number of stars which are circumpolar to and the number which cross the prime vertical of an observer depends primarily upon his latitude. To investigate these problems we shall first consider what may justifiably be regarded as one of the important principles of astronomical navigation, namely:

Latitude of an observer = Altitude of the celestial pole

Fig. 5 serves to prove this important fact. Fig. 5 represents the celestial sphere projected on to the plane of an observer's celestial meridian. The small circle represents the Earth at the centre of the celestial sphere. p represents the Earth's North Pole and qq_1 represents the equator. o represents the observer whose zenith is projected at Z. P is the projection of the elevated celestial pole and QQ_1 is that of the equinoctial. NS lies in the plane of the observer's horizon.

$$\text{arc } qo = \text{arc } QZ = \text{observer's latitude}$$
$$\text{arc } QP = \text{arc } NZ = 90°$$

Therefore:

$$\text{arc } NP = \text{arc } QZ$$

But

$$\text{arc } NP = \text{altitude of celestial pole}$$

Therefore:

$$\text{Latitude of observer} = \text{Altitude of celestial pole}$$

Reference back to Fig. 4 will verify the fact that for a celestial body to be circumpolar its polar distance, which is the complement of its declination, must be smaller than the observer's

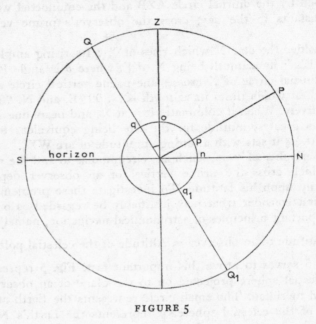

FIGURE 5

latitude; and, of course the names of the latitude and declination must be the same.

Also from Fig. 4 it may be verified that for a celestial body to cross the prime vertical circle of an observer its declination must be smaller than, and of the same name as, the latitude.

A celestial body having a declination of 0° will culminate with an altitude equal to the complement of the observer's altitude. One having a declination the same in name and equal in magnitude to the observer's latitude will culminate at the zenith. In other words, its maximum altitude for the day will be 90°. These facts are readily verified from Fig. 4.

TWILIGHT

Because of a phenomenon known as *atmospheric refraction* (see Part II, Chapter II), and because sunlight may be reflected from particles high in the atmosphere, sunlight may be received by an observer when the Sun is as much as 18° below his horizon. Sunlight received after sunset or before sunrise is called *twilight*.

When the Sun is below and within 6° of the horizon, twilight is sufficiently strong to obviate the necessity, except for exceptional cases, of artificial lighting. This is called *civil twilight*. When the Sun is between about 6° and 12° below the horizon, twilight is sufficiently strong for the seaman to see his visible horizon, but sufficiently weak for many of the brighter stars to be visible. This provides the best conditions for observing star altitudes, and the twilight received in these circumstances is called *nautical twilight*. When the Sun is between about 12° and 18° below the horizon twilight is relatively weak and is called *astronomical twilight*.

The duration of twilight is related to the angle which the plane of the Sun's diurnal circle makes with the plane of the horizon. The bigger is this angle the shorter will be the duration of twilight. At the equator the Sun sinks into and rises out of the horizon perpendicularly, so that the duration of twilight is relatively short. In high latitudes, the Sun's apparent diurnal path at rising or setting makes a relatively small angle (an angle which decreases as latitude increases) with the horizon so that the duration of twilight is relatively long. If the Sun does not sink lower than about 18° below the horizon, twilight will last all night. Fig. 6 serves to illustrate that for twilight to last all night the latitude of the observer and the Sun's declination must have the same name and their sum must be not less than 72°.

Fig. 6 represents the celestial sphere projected on to the plane of the horizon of an observer whose zenith is projected at Z.

N, E, S and W, are the projections of the cardinal points of the horizon. The outer circle represents the parallel of altitude of 18° below the horizon. For twilight to last all night the Sun's diurnal circle must not cross this parallel of altitude. In Fig. 6 the circle centred at P just grazes the parallel of altitude of 18° below the horizon at X. This is the diurnal circle of a celestial

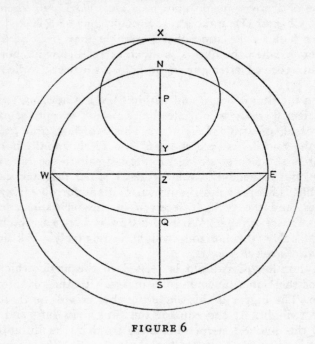

FIGURE 6

body which crosses the observer's celestial meridian when at upper transit at Y. Its declination, therefore, is equal to arc QY.

Now $\qquad$ PX = NP + 18°

that is $\qquad$ PY = NP + 18°

But $\qquad$ PY = 90° − QY

where QY is the declination of the celestial body whose diurnal circle passes through X and Y.

Therefore:

$$NP + 18° = 90° - QY$$

or $$Lat + 18° = 90° - Dec$$

from which

$$(Lat + Dec) = 72°$$

Thus, for twilight to last all night the sum of the observer's latitude and the Sun's declination must equal or be not less than 72°. In other words, if the observer's latitude is not less than (72° − Sun's declination) twilight will last all night. Therefore, the lowest latitude at which twilight can last all night is $(72 - 23\frac{1}{2})°$; that is, latitude $48\frac{1}{2}°$.

CHAPTER VI

Time

Time in the astronomical sense denotes that which persists while astronomical events, such as eclipses, culminations, occultations, Sun's arrivals at the equinoctial and solstitial points, etc., take place. The common units of time are related to astronomical periods: in particular the period of the Earth's rotation; that of the Moon's revolution around the Earth; and the period of the Sun's annual apparent motion on the celestial sphere (which is a reflection of the Earth's real orbital motion around the Sun).

The Earth's period of rotation, although not quite regular, provides a perfect unit of time for ordinary navigational purposes. The period of the Earth's rotation is a natural unit of time called a *day*.

The Earth's rotation is manifested by the apparent diurnal motion of the celestial sphere; so that the celestial bodies, in rising, culminating and setting, may be regarded as pointers which mark off the hours, minutes and seconds of the day.

A simple definition of the basic unit of time is: 'A day is the interval between successive risings, culminations or settings, of a celestial body.'

It is comparatively difficult to time the rising or setting of a celestial body, but comparatively easy to time its culmination. This is done in an observatory using a *transit instrument* which is a telescope of special design set in the plane of the meridian. Thus, a more satisfactory definition of a day is: 'A day is the interval between successive transits of a celestial body across an observer's upper celestial meridian'.

Because of the movements of heavenly bodies relative to one another, on account of the Earth's rotation and revolution, and because of the real movements of the Moon and planets, a day by this definition varies in length according to which type of heavenly body is used.

If any fixed star is used for the purpose of defining a day, the interval between its successive transits with an observer's upper celestial meridian, is a unit of time called a *sidereal day*. In practice, the First Point of Aries, which is a fixed point on the celestial sphere in the sense that stars are fixed, is used for determining the sidereal day: so that the sidereal day is defined as the interval of time that elapses between two successive transits of the First Point of Aries with an observer's celestial meridian.

A clock which registers sidereal time correctly will indicate 00 hr 00 min 00 sec at the instant when the First Point of Aries bears due north or due south at upper transit.

The sidereal day is a constant unit of time, and may be regarded as being a measure of the time taken for the Earth to rotate exactly 360° on her axis. For navigational purposes the sidereal day is a very important unit of time, but for everyday purposes of civil life it plays no part whatever. The Sun is the most important celestial body for most human activities; and this luminary, therefore, is used as a basis for time-measuring for civil purposes.

The interval of time which elapses between successive lower transits of the Sun is a unit of time called a *solar day*.

The solar day, it must be appreciated, commences when the Sun crosses the *lower* celestial meridian of an observer: so that a clock registering correct solar time will indicate 00 hr 00 min 00 sec at the instant when the Sun is at lower transit. When he is at upper transit, half a solar day will have elapsed since the solar day commenced. Because the day is subdivided into 24 hours, a clock registering solar time will indicate 12 hr 00 min 00 sec when the Sun crosses the upper celestial meridian of an observer. It is for this reason that the time of the Sun's upper meridian passage is called *midday*.

Fig. 1 serves to illustrate that a solar day is longer than a sidereal day.

The period of the Earth's revolution around the Sun is about 365 days; so that, as the Earth moves in her orbit, the Sun appears to move eastwards across the celestial sphere at the rate of $\frac{360}{365}$ or approximately 1° per day.

Referring to Fig. 1, in which o represents an observer at whose meridian, when the Earth is at E_1, the solar time is midday. Let us assume that a particular fixed star happens to be in transit with the Sun at this instant. The next time the star will be on the observer's upper celestial meridian will be after the Earth has rotated exactly 360°; that is when the Earth is at E_2 and the observer is at o_1. At this instant the Sun will lie about 1° to the east of the observer's upper celestial meridian,

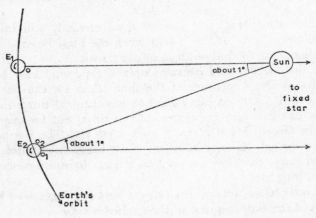

FIGURE I

so that the Sun will not cross this meridian until after the star does. The solar day, therefore, is slightly longer than a sidereal day, and it is the time taken for the Earth to spin about 361° on her axis.

The principal disadvantage of using the solar day, as we have defined it, arises from the variable motion of the Sun in his annual apparent orbit. When the Earth is near perihelion the apparent angular motion of the Sun in the ecliptic is more rapid than it is when the Earth is near aphelion: so that the interval of time between successive transits of the Sun varies with the time of year, being relatively long when the Earth is near perihelion and relatively short when the Earth is near aphelion.

Another factor which influences the length of the solar day is due to the plane of the Earth's spin not being coincident with that of her orbit around the Sun.

To overcome the variations in the length of the solar day due to the combination of the effects of the varying speed of the Earth's orbital motion and the obliquity of the ecliptic, and yet use the Sun as the basis of time measuring, an imaginary point known as the *Mean Sun* is employed. The Mean Sun is a point which moves in the equinoctial at a uniform rate.

The interval of time elapsing between successive upper transits of the Mean Sun is a unit of time called a *Mean Solar Day*. The day by the actual Sun is usually called an *Apparent Solar Day*, because the apparent diurnal motion of the True Sun is used in its determination.

Having described the principal units of time, it is now necessary to understand the meaning of the term 'time at a given instant'. Time at any instant is a measure of an angle swept out by a semi-great circle which is centred at the celestial pole, and which swings with diurnal motion around the sky making one rotation in a day. For indicating sidereal time the semi-great circle referred to is that on which the First Point of Aries is located. For indicating solar time it is that on which the Sun is located—the True Sun for Apparent Solar Time, and the Mean Sun for Mean Solar Time.

The semi-great circles referred to above may be imagined to 'sweep out time'. For this reason they are called *hour circles*: so that local time at any instant may be defined generally as the angle at the celestial pole contained between the observer's meridian (the upper celestial meridian for sidereal time and the lower for solar time) and the hour circle of the celestial point or body used for indicating time, measured westwards from the observer's celestial meridian.

The angle at any instant at the celestial pole contained between the upper celestial meridian of an observer and the hour circle of a celestial point or body, measured westwards from the observer's upper celestial meridian, is known as the *Local Hour Angle* (L.H.A.) of the point or body at that instant.

An hour angle measured westwards from the upper celestial

meridian of Greenwich is called a *Greenwich Hour Angle* (G.H.A.).

The *Local Sidereal Time* (L.S.T.) at any instant is equivalent to the L.H.A. of the First Point of Aries, i.e.

$$\text{L.S.T.} = \text{L.H.A.} \ \Upsilon$$

The *Local Apparent Solar Time* (L.A.T.) is equivalent to the L.H.A. of the True Sun ± 12 hours, i.e.

$$\text{L.A.T.} = \text{L.H.A.T.S.} \pm 12 \text{ hours}$$

The *Local Mean Solar Time* (L.M.T.) is equivalent to the L.H.A. of the Mean Sun ± 12 hours, i.e.

$$\text{L.M.T.} = \text{L.H.A.M.S.} \pm 12 \text{ hours}$$

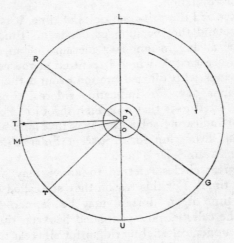

FIGURE 2

Fig. 2 illustrates the celestial sphere, with the Earth at the centre, projected on to the plane of the equinoctial.

o represents an observer and P is the projection of the north celestial pole. PU and PL are the projections of the observer's upper and lower celestial meridians respectively. PG and PR are the projections of the Greenwich upper and lower celestial meridians respectively.

P♈, PM, PT are the projections of the hour circles through the First Point of Aries, the Mean Sun and the True Sun respectively.

$$\text{arc } U♈ = \text{L.H.A. } ♈ \quad = \text{L.S.T.}$$
$$\text{arc } G♈ = \text{G.H.A. } ♈ \quad = \text{G.S.T.}$$
$$\text{arc } UM = \text{L.H.A.M.S.}$$
$$\text{arc } LUM = \text{L.M.T.} \quad = \text{L.H.A.M.S.} + 12 \text{ hour}$$
$$\text{arc } UT = \text{L.H.A.T.S.}$$
$$\text{arc } LUT = \text{L.A.T.} \quad = \text{L.H.A.T.S.} + 12 \text{ hour}$$
$$\text{arc } GM = \text{G.H.A.M.S.}$$
$$\text{arc } RGM = \text{G.M.T.} \quad = \text{G.H.A.M.S.} + 12 \text{ hour}$$
$$\text{arc } GT = \text{G.H.A.T.S.}$$
$$\text{arc } RGT = \text{G.A.T.} \quad = \text{G.H.A.T.S.} + 12 \text{ hour}$$

In Fig. 2 the Mean and True Suns are shown on the western side of the observer's upper celestial meridian. Had they been on the eastern side L.H.A.T.S. and L.H.A.M.S. would have been greater than L.A.T. and L.M.T. respectively, by 12 hours. That is:

$$\text{L.A.T.} = \text{L.H.A.T.S.} - 12 \text{ hours}$$
$$\text{L.M.T.} = \text{L.H.A.M.S.} - 12 \text{ hours}$$

Fig. 2 is an example of what has become known as a *time diagram*, a device which provides a simple method of illustrating and verifying problems related to time and hour angle.

THE EQUATION OF TIME

The Mean and True Suns seldom occupy the same hour circle. At some periods in the years the Mean Sun is *ahead* or west of the True Sun, by which we mean that the Mean Sun's hour angle at any instant exceeds that of the True Sun at the same instant. At other periods the Mean Sun is *behind* or east of the True Sun.

The angle at the celestial pole contained between the hour circles of the Mean and True Suns is called the *Equation of Time* (*E*). When the Mean Sun is ahead or west of the True

Sun E is conventionally named plus ($+$). When the Mean Sun is astern or east of the True Sun it is named minus ($-$): so that E is usually defined as being the excess of Mean Time over Apparent Time. For example, if the L.M.T. is 1050 at the instant when L.A.T. is 1040, the equation of time is described as being $+10$ minutes. If, on the other hand, L.M.T. is 1040 at the instant when L.A.T. is 1050, the equation of time is described as -10 minutes.

$$E = \text{L.M.T.} - \text{L.A.T.}$$

or $$E = \text{L.H.A.M.S.} - \text{L.H.A.T.S.}$$

or $$E = \text{G.H.A.M.S.} - \text{G.H.A.T.S.}$$

or $$E = \text{R.A.T.S.} - \text{R.A.M.S.}$$

In Fig. 2, because the Mean Sun lies to the east of the True Sun, E, which is denoted by arc MT, is a negative quantity.

The equation of time is considered to be composed of two parts: one resulting from the ellipticity of the Earth's orbit around the Sun; and the other resulting from the obliquity of the ecliptic.

An imaginary point which moves in the ecliptic at a uniform rate equal to the average rate of the True Sun's motion is called the *Dynamical Mean Sun* (D.M.S.). That part of the equation of time due to ellipticity is a measure of the angular difference between the hour circles of the True Sun and the D.M.S. It is equivalent to the difference between the R.A.'s of the True Sun and the Dynamical Mean Sun.

The D.M.S. and the True Sun are considered to be in coincidence at perihelion and aphelion (approximately January 3rd and July 3rd). For three months after the time of perihelion the True Sun increases his R.A. at a greater rate than does the D.M.S.; so that the maximum angular separation occurs at about April 3rd. From April 3rd until the time of aphelion their separation diminishes until they coincide at the time of aphelion. For the three months following aphelion the D.M.S. increases its R.A. at a rate faster than that of the True Sun, so that the maximum separation occurs at about October 3rd. For the following three months their separation decreases until they coincide again at the time of the next perihelion.

The maximum angle between the hour circles of the True Sun and the D.M.S. occurs at about April 3rd and October 3rd; and it amounts to about 2° or 8 minutes of time. Between January 3rd and July 3rd, when the R.A. of the True Sun is greater than that of the D.M.S. the hour angle of the D.M.S. at any instant is greater than that of the True Sun, so that the component of the equation of time due to ellipticity is named plus (+). From July 3rd to the following January 3rd it is named minus (−).

The D.M.S. increases its celestial longitude at a uniform rate, so that if the planes of the Earth's rotation on her polar axis and her revolution around the Sun were coincident the D.M.S. would provide the means of regular timekeeping. Because of the obliquity of the ecliptic the perfect astronomical time-keeper must move on the equinoctial at a uniform rate. This is the imaginary point we have called the Mean Sun. To distinguish the Mean Sun from the Dynamical Mean Sun, it is called the *Astronomical Mean Sun* (A.M.S.).

The A.M.S. moves uniformly in the equinoctial increasing its R.A. at the same rate as the D.M.S. increases its celestial longitude.

The component of the equation of time due to obliquity is the difference between the hour angles or R.A.'s of the D.M.S. and the A.M.S.

The A.M.S. coincides with the D.M.S. at the First Point of Aries on March 21st. For the following three months the R.A. of the A.M.S. is greater than that of the D.M.S. This is illustrated in Fig. 3, in which arc ♈D is equivalent to arc ♈A, D and A representing the D.M.S. and the A.M.S. respectively.

The H.A. of the A.M.S. at any instant during northern Spring is less than that of the D.M.S.; so that, during this season, the component of the equation of time due to obliquity is a negative quantity.

Both the A.M.S. and the D.M.S. travel through equal arcs along their respective paths—equinoctial and ecliptic respectively—in equal time intervals; so that at the time of the Summer solstice they occupy the same celestial meridian, and their R.A's are each 06 hours. From the time of the Summer solstice to that of the Autumnal equinox the R.A. of the A.M.S. is less

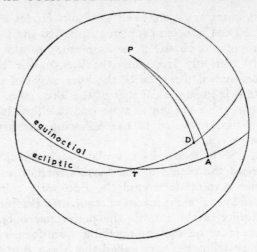

FIGURE 3

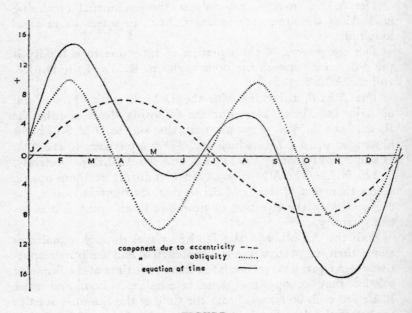

FIGURE 4

than that of the D.M.S., so that the component of the equation of time due to obliquity is plus (+). Similarly, from the time of the Autumnal equinox to that of the Winter solstice it is minus (−); and from the time of the Winter solstice to that of the following Spring equinox it is plus (+).

Fig. 4 illustrates the variations in the components of the equation of time, and also those of the resultant of the two components, which is the equation of time.

COMPARISON OF THE LENGTHS OF THE SIDEREAL AND MEAN SOLAR DAYS

During one revolution of her orbit the Earth makes about 365¼ revolutions on her polar axis. During the Earth's period of revolution the Sun describes one apparent annual circuit of the ecliptic; and, in so doing, makes one apparent revolution relative to the fixed stars. Thus, relative to the fixed stars, the Earth makes one more rotation in a year than she makes relative to the Sun. It follows, therefore, that:

$$365\tfrac{1}{4} \text{ Mean Solar days} = 366\tfrac{1}{4} \text{ sidereal days}$$

From this relationship:

1 Mean Solar day   = 24 hr 00 min 00 sec of Mean Solar Time

                        = 24 hr 03 min 56·5 sec of Sidereal Time

1 sidereal day   = 24 hr 00 min 00 sec of Sidereal Time

                        = 23 hr 56 min 04·1 sec of Mean Solar Time

The solar day, therefore, is about four minutes longer than a sidereal day. For this reason, the time of transit of any fixed star with the upper celestial meridian of any terrestrial observer is about four minutes later on successive days. For this reason the aspect of the heavens changes gradually throughout the year, and different constellations cross an observer's celestial meridian at any given time of night, at different times of the year.

On the day of the Spring equinox, when the Sun's R.A. is 00 hours, celestial objects having equal or nearly equal R.A's will cross the observer's celestial meridian at more or less the same time as the Sun. These objects, therefore, will not be suitably placed in the sky for altitude observations. On the other hand, celestial objects whose R.A's differ by 12 hours from that of the Sun's, will cross the observer's upper celestial meridian when the Sun is on the lower celestial meridian. These bodies will, therefore, cross the observer's meridian at or about midnight. They will be east of the meridian before midnight and west of the meridian after midnight. They may, therefore, be suitably placed for altitude observations during evening twilight when they are east of the meridian, and during morning twilight when they are west.

By comparing the Sun's R.A. (or S.H.A.) with that of any given star, planet or the Moon, a navigator may readily ascertain, after considering the declination of the body, whether it is suitable or not for navigational purposes at his particular location on the globe.

LONGITUDE AND TIME

The angle at the celestial pole contained between the upper celestial meridians of Greenwich and any observer is equivalent

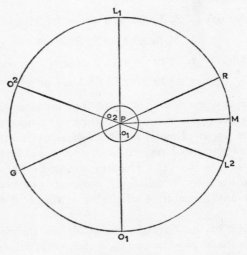

FIGURE 5

to the longitude of the observer. This follows from the fact that the Greenwich meridian, from which longitudes are measured, lies in the same plane as the Greenwich upper celestial meridian; and the observer's terrestrial meridian lies in the same plane as the observer's upper celestial meridian.

The problem of finding longitude by astronomical methods is, therefore, essentially a problem of comparing the local time of an astronomical event (usually the instant when a heavenly body has a particular observed altitude) with the Greenwich time of the same instant.

Fig. 5 illustrates the celestial sphere drawn on the plane of the equinoctial with the Earth at the centre. o_1 and o_2 are observers located in the eastern and western hemispheres respectively. G represents Greenwich. P is the projection of the north celestial pole.

PO_1 represents o_1's upper celestial meridian

PL_1 represents o_1's lower celestial meridian

PO_2 represents o_2's upper celestial meridian

PL_2 represents o_2's lower celestial meridian

PG represents the Greenwich upper celestial meridian

PM represents the hour circle of the Mean Sun

$$\text{arc RM} \qquad = \text{G.M.T.}$$
$$\text{arc } L_1M \qquad = \text{L.M.T. at } o_1$$
$$\text{arc } L_2O_2M = \text{L.M.T. at } o_2$$
$$\text{arc } GO_1 \qquad = \text{East longitude of } o_1$$
$$\text{arc } GO_2 \qquad = \text{West longitude of } o_2$$

Now:

$$\text{arc } GO_2 = \text{arc } RL_2$$
$$= \text{arc RM} + \text{arc } ML_2$$
$$= \text{arc RM} + (24 \text{ hr} - \text{arc } L_2O_2M)$$
$$= \text{arc RM} - \text{arc } L_2O_2M$$

Therefore:

$$\text{West longitude of } o_2 = \text{G.M.T.} - \text{L.M.T. at } o_2$$

Also:

$$\text{arc } GO_1 = \text{arc } L_1R$$
$$= \text{arc } L_1M - \text{arc } RM$$

Therefore:

East longitude of o_1 = L.M.T. at o_1 − G.M.T.

It follows, therefore, that the longitude of a terrestrial position is a measure of the difference between L.M.T. and G.M.T. at any instant, reckoned at the rate of 15° of longitude per one hour difference between L.M.T. and G.M.T.

If G.M.T. is greater than L.M.T., longitude is named west. If it is less than L.M.T., longitude is named east. Hence the well known seaman's rhyme:

'Longitude west, Greenwich time best;
Longitude east, Greenwich time least'.

TIMEKEEPING AT SEA

It is impracticable, if not impossible, to keep local time on a moving ship, unless her course is along a meridian. Local time is a measure of an angle at the celestial pole between an observer's lower celestial meridian and the hour circle through the Sun. Because of the movement of the observer's lower celestial meridian, as a result of his easterly or westerly motion on the Earth's surface, it would be necessary, if the observer wished to keep local time, for him continually to alter his clock time at a rate proportional to his motion in longitude.

In days gone by it was the custom in merchant ships to set the clock at 12 o'clock when the Sun reached his greatest altitude of the day. About a quarter of an hour or so before noon, the Captain and navigating officers would assemble on poop or bridge, armed with their sextants to observe the changing altitude of the Sun. When this luminary reached his greatest altitude at his upper meridian passage, his L.H.A. would be 00 hours, and the solar time would, accordingly, be 12 hr 00 min or midday. At the time of meridian passage, therefore, the ship's clocks would be set to 12 o'clock and the order given for eight bells to be made. During the second dog watch, at about seven in the evening, the navigator would estimate the ship's longitude for the following apparent noon. This enabled him to find the error of the clock on the apparent time for the meridian at which he estimated his ship would be at the next apparent noon. The ship's clocks would then be altered, or *flogged* as the seamen say, in the hope that at the following apparent noon they would register 12 o'clock. If this hope was fulfilled the clock afforded a reliable guide to the time of the Sun's upper meridian passage on the following day.

This system of time organization is still used to some extent, although it appears to be giving way to the system of *zone time*.

In the zone time system the Earth is divided into north–south parts called *time zones*: these (with two exceptions) being bounded by meridians whose d.long is 15° or one hour in time. Each time zone, of which there are twelve in the western and twelve in the eastern hemisphere, is designated by a zone number which is pre-fixed by a plus (+) sign for those in the western hemisphere and a minus (−) sign for those in the eastern hemisphere.

The zone time (Z.T.) in any given zone is always an integral number of hours different from G.M.T., the number being the same as the zone number. If the zone number is (+) the G.M.T. at any instant is equal to the Z.T. plus a number of hours equal to the zone number. If the zone number is (−) the G.M.T. at any instant is equal to the Z.T. minus a number of hours equal to the zone number. Thus, if it is, say, 1020 Z.T. in zone (+4) it is 1420 G.M.T. If it is, say, 1650 Z.T. in zone (−6) it is 1050 G.M.T., etc.

Zone +1 extends from $7\frac{1}{2}°$ W. to $22\frac{1}{2}°$ W.; and all ships keeping Z.T. in zone +1 keep time which is one hour astern of G.M.T. Zone −1 extends from $7\frac{1}{2}°$ E. to $22\frac{1}{2}°$ E.: and all ships keeping zone time in zone −1 have their clocks set one hour ahead of G.M.T. The zone between $7\frac{1}{2}°$ E. and $7\frac{1}{2}°$ W. is called zone 0; and all ships keeping zone time within this zone have their clocks set to G.M.T.

On crossing the boundary of a time zone the clock is altered one hour abruptly; so that when sailing westwards the clock is retarded and when sailing eastwards it is advanced.

The 15° zone antipodal to zone 0 is bisected longitudinally by the 180th meridian. That half lying between 172½° W. and 180° is designated zone +12. The other half is designated zone −12. When crossing the 180th meridian the zone number changes from +12 to −12 when sailing westwards, and from −12 to +12 when sailing eastwards. It follows, therefore, that when crossing the 180th meridian the date will have to be changed, advancing the date by a day when sailing westwards, and retarding it a day when sailing eastwards. It is for this reason that the 180th meridian is called the *date line*.

To overcome the difficulties arising from keeping local time at places ashore, it was long ago agreed internationally that shore time should be systematized rationally. The zone time system is admirably suitable for this purpose; and it is used extensively ashore, as well as at sea. For shore purposes, the boundaries of time zones are sometimes adjusted to take in national territory outside the normal boundaries of the time zones. The date line, for example, does not coincide exactly with the 180th meridian. It is adjusted in order that certain Pacific islands and other territory which straddles the 180th meridian, which have a common administration, keep a common time. The times kept by nations are called *standard times*. They are usually related to G.M.T. Information about standard times is to be found in the *Nautical Almanac*.

YEARS

The period of revolution of the Earth around the Sun provides a natural unit of time called a *year*. The time taken for the Earth to make one revolution relative to any fixed celestial point is called a *sidereal year*. It is 365 days 06 hr 09 min 09 sec of Mean Solar Time.

Because of the precession of the equinoxes the time taken for the Mean Sun to move from the First Point of Aries back to the same point is slightly shorter than a sidereal year. It is 365 days 05 hr 48 min 46 sec of Mean Solar Time. It corresponds to the interval between successive Spring equinoxes and is the period of the seasons. It is called a *tropical year*.

The straight line joining perihelion and aphelion, a line called the *apse line* or *line of apsides*, is not fixed in the plane of the Earth's orbit: it swings slowly around the Earth's orbit in such a way that the dates of perihelion and aphelion occur progressively later. The interval between successive perihelions is slightly longer than a sidereal year. The interval, which is called an *anomalistic year*, is 365 days 06 hr 13 min 48 sec of Mean Solar Time.

THE CALENDAR

The systematic arrangement of units of time constitutes a calendar. The incommensurable nature of the natural units of time, viz. the day, month and year, made the problem of fitting them together in an orderly way one of great difficulty to the astronomers of old.

An early attempt is that known as the *Julian calendar* named in honour of Julius Caesar and contrived by the Alexandrian astronomer Sosigenes. The year, by the Julian reckoning, is 365 days 6 hours exactly. The 365 days were divided into twelve months each containing an integral, but not necessarily the same, number of days. The extra six hours in the year were allowed to accumulate for four years making an extra day which was intercalated to form a year containing 366 days instead of 365 days as in the common or ordinary years. The intercalated day was called the *bissextus*, and a year which contained it a *bissextile year*.

The year that is of the greatest significance in calendar making is the tropical year, this regulating, as it does, the seasons. By taking the year as 365 days 6 hours an error amounting to about 11 minutes a year throws out the calendar according to the recurring seasons. The Julian calendar was used in Britain until 1752, by which date the accumulated error amounted to 11 days.

The *Gregorian calendar* was introduced in Britain in 1752. This calendar takes it name from Pope Gregory XIII who occupied the Papal throne during part of the 16th century. Pope Gregory's calendar took into account the error in the Julian calendar, which amounts to very nearly 72 hours or three days in 400 years. In the Gregorian calendar three bissextile or leap years are dropped every 400 years, these being the opening years of centuries except those in which the first two numbers of the year is divisible by 4. Thus the year 2000 will be a leap year whereas the year 1900 was not.

Before the Julian calendar was introduced the Romans employed a calendar in which the period of the Moon's revolution around the Earth played the principal element. The months commenced on the days of the New Moon. These days were the *calends* of the months. The middle day of each month, the day at which the Moon is at the Second Quarter, is called the *ides* of the month.

The Moon's motion is still used in the ecclesiastical calendar from which the dates of the moveable feasts of the church are derived. The principal feast day in the Christian year is Easter Day, from which all the other moveable feasts are found. In general, Easter Day falls on the first Sunday after the Full Moon which follows the Spring equinox. The date of the Full Moon, for this purpose, is calculated from an eclipse cycle known as the *Metonic Cycle*—named after Meton of Athens who first discovered it. The Cycle of Meton is a 19-year period; and each year in the cycle is given a *Golden Number* between 1 and 19. The Golden Number is found by adding one to the year number and dividing the result by 19. The remainder is the Golden Number. This is used in conjunction with the *Sunday* or *Dominical Letter*, which is the letter for Sunday starting with A for the first day of the year. If, for example, the first day of the year is a Tuesday the Sunday letter for that year will be F; if it falls on a Saturday it will be B, etc.

The subject of calendar making, however interesting it may be, must be brought to an end to make way for the principal parts of our main subject of nautical astronomy. These will be found in the following chapters.

PART II

The Theory of Nautical Astronomy

CHAPTER I

The Astronomical Triangle

The fundamental feature of nautical astronomy, in the modern sense, is the relating of the celestial position of a heavenly body at a given instant of time using the horizon system, with its position at the same instant using the equinoctial system. The co-ordinates employed in defining a celestial position using the horizon system are altitude and azimuth; and those employed for navigational purposes in defining a celestial position using the equinoctial system are declination and hour angle.

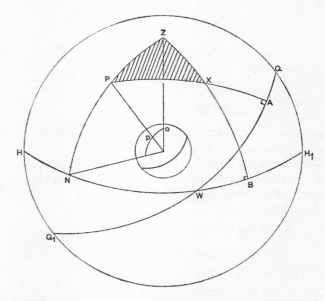

FIGURE I

81

The hour angle and azimuth of an observed celestial body, together with a third angle called the *parallactic angle* or the *angle of position*, form the three angles of the celestial spherical triangle which is the subject of this chapter. Two of the sides of the astronomical triangle are functions of the altitude and declination of the observed celestial body. The third side is a function of the observer's latitude.

Fig. 1 illustrates a typical astronomical triangle.

Fig. 1 illustrates the celestial sphere with the Earth at its centre. o is an observer in the northern hemisphere, and Z is his zenith. p is the Earth's North Pole and P is the elevated pole. HH_1 is the observer's horizon and QQ_1 is the equinoctial. ZPN is the vertical circle through the north point N of the observer's horizon. This, therefore, is the observer's celestial meridian. W is the west point of the observer's horizon, and X is a celestial body.

The spherical triangle PZX is the astronomical triangle related to the observer o and the celestial body X.

$$\text{arc XA} = \text{declination of X}$$
$$\text{angle ZPX} = \text{L.H.A. of X}$$

These two co-ordinates define the body's position using the equinoctial system.

$$\text{arc XB} = \text{altitude of X}$$
$$\text{angle PZX} = \text{azimuth of X}$$

These two co-ordinates define the body's position using the horizon system.

We have seen, in Part I, Chapter 5, that the altitude of the celestial pole is equivalent to the observer's latitude, so that in Fig. 1 arc NP is equal to the observer's latitude.

In the PZX triangle:

$$PZ = (90° - NP)$$
$$= \text{co-latitude of observer}$$
$$ZX = (90° - BX)$$
$$= \text{co-altitude of observed body}$$
$$= \text{zenith distance of observed body}$$

PX $=$ (90° − XA)

$\qquad$ = co-declination of observed body

(N.B. Had the declination of the observed body been south the arc PX would have been greater than 90° by an amount equal to the declination.) In all cases

$\qquad$ PX = polar distance of observed body

$\qquad\qquad$ = (90° − declination) when latitude and declination have the same name

$\qquad\qquad$ = (90° + declination) when latitude and declination have different names

angle ZPX = L.H.A. of observed body

angle PZX = azimuth of observed body

angle ZXP = parallactic angle (this plays only a minor role in nautical astronomy)

If the observer knows his latitude (and this is not generally the case with a navigator) the angle P of the PZX triangle may be computed using the three sides of the astronomical triangle.

The angle P is the local hour angle (L.H.A.) of the observed celestial body. This, when compared with the Greenwich hour angle (G.H.A.) of the body at the time of the observation, will yield the observer's longitude. The G.H.A. is provided in the *Nautical Almanac* against G.M.T., so that an essential process in nautical astronomy is timing an altitude observation of a heavenly body, using a chronometer the error on G.M.T. of which is known.

$\qquad$ G.H.A. of * $\sim$ L.H.A. of * = Longitude of observer

where * is any celestial body.

In general, an astronomical triangle is solved by the navigator when he wishes to obtain an astronomical position line. A *position line* is a line drawn on a navigational chart somewhere on which the navigator may fix his ship's position. Notice that the result of a PZX triangle computation is a *line* of position and not a *point* of position.

The principal and significant difference between finding position ashore and at sea is that for the shore station which is fixed, the latitude of the station is first found. This, as we shall see in Chapter IV, is a relatively simple matter. Having found the latitude, it is used in a PZX triangle for the purpose of finding the longitude of the station. The sea observer, being on a moving ship, is never certain of his latitude at the time he observes to find his longitude, so that he has to use an estimated latitude in order to form an astronomical triangle. If the latitude used in forming this triangle is, in fact, the ship's actual latitude, the ship's longitude may be found without difficulty. If, on the other hand, the latitude used is not the ship's latitude, the ship's longitude ascertained from the astronomical triangle will be in error proportional to the difference between the estimated and actual latitudes of the ship.

It is possible, as we presently shall see, to obtain a position line from an observation or 'sight' of a heavenly body, in spite of the fact that the latitude used in computing the L.H.A. of the observed body is not the ship's actual latitude. A position line obtained from a celestial observation is called an *astronomical position line*. An astronomical position line is the projection on a chart of part of a *circle of equal altitude* the centre of which is located at a point on the Earth called the *geographical position* of the observed body. Let us discuss this type of circle and its centre in some detail.

THE GEOGRAPHICAL POSITION OF A HEAVENLY BODY

The geographical position (G.P.) of a heavenly body at a given instant of time is simply a point on the Earth at which the body is in the zenith at the instant. If the body referred to is the Sun, his G.P. is called the *sub-solar point*. If it is a star, the term *sub-stellar* point is used.

If an observer were to observe a celestial body at his zenith, his terrestrial position would coincide with the observed body's G.P. at the time of the observation. If, therefore, the observer knows the G.P. of the body at the time of a zenithal observation, his own position would also be known, this coinciding with the body's G.P. This interesting navigational sight provides the only case whereby a ship's position may be found from a solitary observation. This celestial observation forms the basis of primitive Polynesian navigation by means of which the intrepid voyagers of the South Seas were able to make long sea journeys with navigational precision.

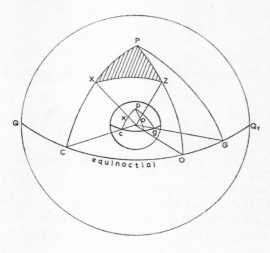

FIGURE 2

Fig. 2 serves to illustrate how the latitude and longitude of the G.P. of a heavenly body at any given instant are related to the declination and G.H.A. of the body at the time of the observation.

In Fig. 2, QQ_1 represents the equinoctial and P the north celestial pole. p is the Earth's North Pole, o is an observer and Z is his zenith. pg represents the Greenwich meridian. PO and PG are the upper celestial meridians of the observer and Greenwich respectively. PC is the hour circle of the celestial body X.

The point x is the G.P. of the celestial body X.

$$\text{Latitude of G.P. of } X = \text{arc } cx$$

$$\text{Declination of } X = \text{arc } CX$$

But

$$\text{arc } cx = \text{arc } CX$$

therefore:

Latitude of G.P. of X = Declination of X

Longitude of G.P. of X = angle gpx

G.H.A. of X = angle GPX

But   angle gpx = angle GPX

therefore:

Longitude of G.P. of X = G.H.A. of X

Now,   G.H.A. of X = GPX

= GPO + ZPX

therefore:

G.H.A. of X = L.H.A. of X + W. longitude of observer

The *Nautical Almanac* provides data from which the declination and G.H.A. of any celestial body of navigational importance for any given G.M.T. may be found. Thus, if G.M.T. is known at any instant the G.P. of any navigational body for that G.M.T. may be found.

For the Sun, Moon and navigational planets, the *Nautical Almanac* gives values of G.H.A. and declination for integral hours of G.M.T. for the whole year. Interpolation tables are provided so that the G.H.A. and declination of any of these bodies may be found for any G.M.T. other than an integral hour.

In order to find the longitude of the G.P. of any fixed star for any given G.M.T., the star's G.H.A. is found by adding its S.H.A. to the G.H.A. ♈; which latter, like the G.H.A. of Sun, Moon or planets, is tabulated for every integral hour of G.M.T. Fig. 3 illustrates that:

G.H.A. of * = G.H.A. ♈ + S.H.A. of *

Fig. 3 illustrates the celestial sphere on the plane of the equinoctial. P is the projection of the celestial pole of the north celestial hemisphere. PG, PX and P♈ are the projections of the celestial meridians through Greenwich, and star *, and the First point of Aries, respectively.

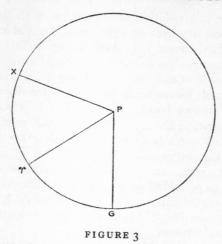

FIGURE 3

From Fig. 3:

$$\text{arc } GX = \text{arc } G \Upsilon + \text{arc } \Upsilon X$$

therefore:

$$\text{G.H.A. of } * = \text{G.H.A. } \Upsilon + \text{S.H.A. of } *$$

The interpolation tables provided in the *Nautical Almanac* for the purpose of finding the G.H.A. of a navigational celestial body give increments to the G.H.A. of Sun, Moon and the First Point of Aries, for every minute and second between 00 min 00 sec and 60 min 00 sec.

The hour angle of the Mean Sun increases at a uniform rate of 15° per hour. The hour angle of the True Sun increases irregularly, but the variation from the mean rate of increase (which is the same as that of the Mean Sun) is so small that the interpolation tables for use with the Sun are devised on the assumption that the True Sun's angle increases uniformly at the rate of 15° per hour exactly. To eliminate any slight error that would arise from this assumption, the tabulated values of the Sun's G.H.A. for integral hours of G.M.T. are adjusted where necessary.

The hour angle of the First Point of Aries increases at a uniform rate of 15° 02·46' per hour, so that the increments

7

lifted from the interpolation tables for use with the First Point of Aries are slightly greater than those for the corresponding increments lifted from the Sun interpolation tables. Care, therefore, must be taken to use the correct interpolation table.

The hour angle of the Moon increases at an erratic rate. The interpolation table for use with the Moon is based on the assumption that the Moon's hour angle increases at a uniform rate of 14° 19′ per hour. This is the value of the minimum rate of change of the Moon's hour angle. To allow for the difference between the minimum and the actual rate of change of the Moon's hour angle, an additional correction called the 'v correction' is to be added to the increment lifted from the main interpolation table. The excess of the Moon's hourly increase in hour angle over 14° 19′ is tabulated as 'v' on the daily pages of the *Nautical Almanac*.

The planets, like the Moon, increase their hour angles at an erratic rate. The interpolation table for the Sun is used for finding the G.H.A. of a navigational planet, the average rate of change of a planet's hour angle seldom departing very much from the Sun's rate of change of hour angle. Sometimes a planet changes its hour angle at a greater rate, and sometimes at a less rate, than that of the Sun, so that the 'v correction' for a planet is sometimes to be subtracted (when its rate of change of hour angle is greater than that of the Sun), and sometimes added (when its rate of change of hour angle is less than that of the Sun).

The interpolation tables provided for finding the G.H.A. of a body are also used for finding the declination of a body for a time which is not an integral number of hours G.M.T. The mean hourly change in the declination of the Sun or planet is tabulated daily, and that for the Moon (whose declination changes relatively rapidly), every hour. This mean hourly change is tabulated as 'd'.

CIRCLES OF EQUAL ALTITUDE

A circle of equal altitude is a circle on the Earth's surface centred at the geographical position of a heavenly body. At every point on a circle of equal altitude the heavenly body has the same altitude. The greater is the altitude of a heavenly body, the smaller is the radius of the circle of equal altitude, the centre of which lies at the geographical position of the observed body. The circle of equal altitude at every point on which the altitude of the body is zero, is a great circle. If the body is the Sun, this circle of equal altitude coincides with the circle of illumination.

Assuming the Earth to be a perfect sphere, the radius of a circle of equal altitude, which is equal to the angle at the Earth's centre between the radius terminating at the G.P. of the object and that terminating at any point on the circle measured in the plane of a great circle through the G.P. of the body, is equivalent to the zenith distance of the body. It follows that the great-circle arc of the Earth's surface in miles between the G.P. of a heavenly body and any point on a circle of equal altitude is equal to the zenith distance of the body in minutes of arc.

An observer standing on a particular circle of equal altitude and facing in the direction of its centre will also be facing in a vertical plane coinciding with that of the vertical circle through the body. In other words, the direction of the G.P. of a heavenly body corresponds to the azimuth of the body at the time of the observation. It follows, therefore, that if the azimuth of the body at the time of the observation can be found, the direction of the circle of equal altitude—which lies at right angles to the direction of any of its radii—can also be found.

The principal problems with which the nautical astronomer is faced, are: finding the position of a point on, or at a known distance from, a circle of equal altitude, and finding the direction of the circle of equal altitude at this point so that he may project part of the circle of equal altitude on his navigational chart or a plotting sheet—this projection being the desired position line. Two such position lines will, if they intersect, provide the navigator with an astronomical fix.

CHAPTER II

The Altitude Corrections

The fundamental process in astronomical navigation is the measuring of the altitude of a celestial body by means of a sextant. We shall discuss the navigator's altitude-measuring instrument in Part III. In this chapter we shall investigate the several corrections that may have to be applied to the altitude measured by means of a perfect sextant in correct adjustment in order to find what is called the *true altitude* of the observed body. This subtracted from 90° gives the *true zenith distance* of the body, an arc which forms one of the three sides of the astronomical triangle PZX.

The true altitude of a celestial body is defined as a measure of the arc of a vertical circle contained between the true direction of the centre of the body at the Earth's centre and the celestial horizon.

The measured altitude is called the *sextant altitude*. If the sextant possesses known error, this is applied in reverse to the sextant altitude to obtain the *observed altitude*.

The true altitude is found by applying *altitude corrections* to the observed altitude. Altitude correction tables are provided in the *Nautical Almanac* as well as in nautical table collections such as those of Burton's and Norie's.

The observed altitude of a celestial body is a measure of the arc of a vertical circle contained at the observer's eye between the apparent direction of the body (or, in the case of the Sun or Moon, the upper or lower limb of the body) and the apparent direction of the visible horizon. The term *apparent direction* is used to denote the fact that the line of sight, which is tangential to the generally curved path through which light from the observed body and the horizon travels, is not coincident with the corresponding true straight-line direction of the body because of a phenomenon known as *refraction*.

The *visible* or *sea horizon* is a small circle on the sea surface which limits the observer's view. It is the line which separates the sea from the sky. Because the observer's eye is elevated above sea level, the sea horizon is depressed below the horizontal plane on which the observer's eye rests. The circle on the celestial sphere which lies in a horizontal plane through the observer's eye is called the observer's *sensible horizon*. The measure of an arc of a vertical circle contained between the sensible and visible horizons is a function of the observer's height of eye above the surface of the sea. It is an angle called the *dip of the sea horizon*.

The great circle on the celestial sphere whose poles are the observer's zenith and nadir respectively is often called the *celestial* or *rational horizon*, to distinguish it from the observer's sensible and visible horizons.

If the observed body is the Sun or Moon the arc of a vertical circle contained between the upper or lower limb of the body and the horizon vertically beneath is measured. The observed altitude must, therefore, be adjusted with an angle which is equivalent to half the angular diameter of the observed body. This is the so-called *semi-diameter correction*.

If the altitude of the relatively close Moon is observed, the fact that her true directions from the Earth's centre and the observer's eye are markedly different necessitates a correction called *parallax*.

In addition to the effects of refraction, height of eye, parallax and semi-diameter, irradiation effect and personal error may influence the altitude of a celestial body. We shall discuss each of these factors in some detail.

REFRACTION

Observations of celestial bodies are made possible through light emitted by or reflected from them. The stars and the Sun are rendered visible by electro-magnetic radiation of optical frequency which is emitted from these astronomical bodies. The Moon and the planets, on the other hand, shine by reflected sunlight.

The path of light is straight only when the light travels through a medium of uniform optical density. When light travels from one medium to another of different optical density, its path direction changes by an angle known as *refraction*.

The air through which light from an observed celestial body or from the horizon travels is not of uniform optical density. It follows, therefore, that the true directions of celestial body and horizon are not the same as those in which they appear to lie. The angular value of the difference between the true and apparent directions of a celestial body is called *celestial* or *astronomical refraction*. That between the true and apparent directions of the horizon is called *terrestrial refraction*.

Celestial and terrestrial refraction depend upon changes in the density of the air along the path taken by the light that enters the observer's eye. Density changes arise from changes in pressure and temperature. These, therefore, are the principal factors which influence refraction.

The law of refraction was first enunciated by the Dutch philosopher Willebrord Snell (1591–1626). Snell's law asserts that when light passes from one medium to another of different density, the planes of the angles of incidence and refraction and the normals to the common surface of the two media are co-incident; and that the sines of the angles of incidence and refraction are in a constant ratio for any two given media. This ratio is called the *refractive index* for the two media.

Several investigations into the phenomenon of celestial refraction were made during the 19th century, and the names of many illustrious astronomers are closely linked with these investigations.

The law of astronomical refraction propounded by Cassini is based on the assumption that the atmosphere is spherical and homogeneous.

In the simplest investigation the Earth is regarded as being flat and the atmosphere is considered to be composed of an infinite number of horizontal parallel layers of air, the density of which decreases uniformly with height above the Earth's surface. On this assumption it is readily proved that the effect of atmospheric refraction is the same as if light entering the atmosphere were refracted directly into the lowest layer of air without traversing the intervening layers.

From Snell's law, a ray of light passes through the atmosphere such that $\mu \sin z$ is constant for every point in its path, μ being the refractive index at any point, and z the angle the path makes with the vertical at the point. If z_0 is the value of z when the ray enters the atmosphere; then, since *in vacuo* the refractive index is unity:

$$\mu \sin z = \sin z_0$$

If μ and z are now taken as referring to the position of an observer's eye, and if r is the atmospheric refraction, then

$$z_0 = z + r$$

Hence $\qquad\qquad \sin z_0 = \sin (z + r)$

i.e. $\qquad\qquad \sin z_0 = \sin z \cos r + \cos z \sin r$

Since r is a small angle (never more than about $\frac{1}{2}°$) we may assume the equivalence of $\sin r$ and r radians, and treat $\cos r$ as 1. We may, therefore, write

$$\mu \sin z = \sin z + r \cos z$$

from which

$$r = (\mu - 1) \tan z$$

i.e. $\qquad\qquad r = U \tan z$

where $U = (\mu - 1)$ or the *coefficient of refraction*.

This result holds good for small zenith distances, but for large zenith distances, by treating $\sin r$ and $\cos r$ as r and 1 respectively, significant error results. Moreover, light from celestial bodies at small altitudes has to travel through a considerable length of atmosphere, and we are not justified, therefore, in regarding the layers of air as being bounded by horizontal parallel planes. Cassini recognized this, and accordingly took into account the Earth's spherical form.

Cassini's formula for atmospheric refraction is explained with reference to Fig. 1.

Fig. 1 represents part of a vertical section through the Earth's centre C and an observer O. XO_1O represents a ray of light from a celestial body X entering an observer's eye.

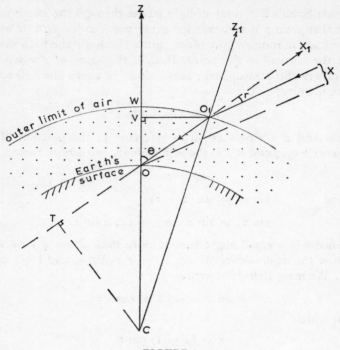

FIGURE I

Cassini's hypothesis is that the light undergoes a single refraction on entering the atmosphere at O_1.

Let the apparent zenith distance of the celestial body be θ, and the true zenith distance θ_1. Let the refraction be r radians.

If r is small,

$$r = (\mu - 1) \tan \theta_1$$

Cassini expressed $\tan \theta_1$ in terms of $\tan \theta$. This he did after first drawing CT perpendicular to O_1O produced; and O_1V perpendicular to COZ. Then:

$$O_1T \tan \theta_1 = OT \tan \theta$$

i.e.

$$\frac{\tan \theta}{\tan \theta_1} = \frac{O_1T}{OT}$$

$$= 1 + \frac{OO_1}{OT}$$

$$= 1 + \frac{OV \sec \theta}{OC \cos \theta}$$

$$= 1 + \frac{OV}{OC} \sec^2 \theta$$

Now OV is approximately equal to the vertical height of the atmosphere OW and is, therefore, equal to $x.OC$, where x is the ratio between the height of the homogeneous atmosphere and the Earth's radius. Therefore:

$$\frac{\tan \theta}{\tan \theta_1} = 1 + x \sec^2 \theta$$

or

$$\tan \theta_1 = \frac{\tan \theta}{1 + x \sec^2 \theta}$$

Expanding the denominator $(1 + x \sec^2 \theta)$ by the binomial theorem, we get:

$$\tan \theta_1 = \tan \theta (1 - x \sec^2 \theta + x^2 \sec^4 \theta - \cdots)$$

Since x is a small quantity, powers of x greater than 1 may be ignored without introducing material error.

Thus:

$$\tan \theta_1 = \tan \theta (1 - x \sec^2 \theta)$$

and

$$r = (\mu - 1) \tan \theta (1 - x \sec^2 \theta)$$

which is Cassini's formula.

If a suitable value for x is chosen, Cassini's formula yields good results for altitudes not less than about 10°.

One of the best of the 18th-century tables of atmospherical refraction was that made by the French astronomer the Abbé de la Caille. He recognized that atmospheric refraction varies with air pressure and temperature, both these factors influencing the air density and, therefore, the refractive index. De la Caille's table gives Mean refractions computed for a standard atmosphere having a specified pressure and temperature at sea level.

Doctor James Bradley, the Astronomer Royal, compiled one of the best refraction tables of the 18th century. Bradley's table consisted of a Mean refraction table calculated for a sea-level pressure and temperature of 29·6 inches of mercury and 50°F respectively, and also an auxiliary table for correcting the Mean refraction when atmospheric conditions differed from those for which the Mean refraction table was based.

It will be of interest to discuss the methods by which atmospheric refraction may be ascertained. The usual method involved the observation of circumpolar stars, and this is explained with reference to Fig. 2.

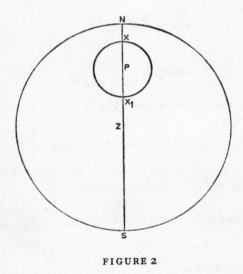

FIGURE 2

Fig. 2 represents the projection of the celestial sphere on to the plane of the horizon of an observer whose zenith is projected at Z. P is the projection of the celestial pole and X and X_1 are the projections of a circumpolar body at lower and upper meridian passage respectively.

Let z and z_1 be the apparent zenith distances of the star when at lower and upper transit respectively. Let p be the polar distance of the star and U the coefficient of refraction ($\mu - 1$),

then:
$$PZ = ZX - PX$$

i.e.
$$PZ = z + U \tan z - p \qquad (1)$$

Similarly:
$$PZ = z_1 + U \tan z_1 + p \qquad (2)$$

Adding (1) and (2):
$$2PZ = z + z_1 + U(\tan z + \tan z_1) \qquad (3)$$

In a like manner, if $\bar{z}$ and $\bar{z}_1$ are the apparent zenith distances of another circumpolar star whose declination differs materially from that of the first star, we have:
$$2PZ = \bar{z} + \bar{z}_1 + U(\tan \bar{z} + \tan \bar{z}_1) \qquad (4)$$

From (3) and (4) we have:
$$z + z_1 + U(\tan z + \tan z_1) = \bar{z} + \bar{z}_1 + U(\tan \bar{z} + \tan \bar{z}_1)$$

from which
$$U = \frac{(z + z_1) \sim (\bar{z} + \bar{z}_1)}{(\tan \bar{z} + \tan \bar{z}_1) \sim (\tan z + \tan z_1)}$$

By repeated observations of circumpolar stars, Bradley found the value of U to be 57·54 seconds of arc.

The value of U used at the present time is usually given for a standard atmosphere of sea-level pressure and temperature of 30 inches of mercury and 50°F, and is 58·3 seconds of arc.

It may be remarked that the actual refraction of light from a heavenly body whose altitude exceeds about 10° is never more than about half a minute of arc different from the Mean refraction. It is for this reason that the auxiliary table to that of Mean refractions is seldom used in practical navigation.

Refraction of light from heavenly bodies within a few degrees of the horizon can never be known with exactitude. Neither the most refined mathematical investigation nor the most careful observations can remove the uncertainty of refraction at small altitudes. Temperature changes—and, therefore, density changes—of the air along the line followed by a ray of light from an object near the horizon, are almost always taking place. These changes can never be known with certainty, and no refraction law has yet been formulated which will hold good at all times for altitudes of less than about 5°.

The value of the refraction for any given atmospheric conditions depends upon the altitude of the body. It varies approximately as the cotangent of the altitude. Its maximum value is about 33' when the altitude is zero. It disappears when the altitude is about 90°, because light from an object in the zenith strikes the observer's eye travelling in a direction normal to the horizontal. The rapid rate of change of refraction with altitude at small altitudes accounts for the oval shape of the Sun and Moon when these bodies lie very near to the horizon.

Abnormal refraction may occur when there is a great difference between air and sea temperatures. If this is suspected, the results of observations of celestial bodies should be used with extreme caution.

DIP OR DEPRESSION OF THE VISIBLE HORIZON

The depression, or dip, of the sea horizon is a measure of the angle contained between the plane of the horizontal surface through the observer's eye, and the direction of the visible horizon. It is an arc of a vertical circle and is a function of the observer's height of eye.

Although it is customary to define the visible horizon as a small circle on the sea surface, the visible horizon is not, in fact, a circle. This follows because of the oblate spheroidal shape of the Earth. The ellipticity of the geoid being a very small fraction means that for practical navigation the Earth may be regarded as being perfectly spherical. On this assumption, the angle of dip is independent of the direction in which an observer may be facing. The dip of the sea horizon is related to the distance of the sea horizon. Fig. 3 illustrates this and serves to demonstrate the relationship between dip, distance of sea horizon and height of observer's eye.

Fig. 3 illustrates the Earth, assumed spherical, and an observer O whose eye is located h feet above the sea. OX is in the plane of the observer's sensible horizon, and OH is a tangent to the Earth's surface at D. C is the centre of the Earth.

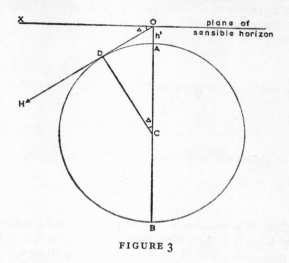

FIGURE 3

The circle on the Earth the radius of which is equal to OD is called the observer's *theoretical* or *geometrical horizon*. From a well-known theorem of plane geometry we have:

$$OD^2 = OA.OB$$

i.e.
$$OD = \sqrt{OA.OB}$$

But, because h is small compared with the Earth's diameter no material error is introduced by assuming that OD is equal to arc length AD and that the Earth's diameter is equal to OB. Therefore:

$$AD = \sqrt{OA.OB}$$

or

$$\text{Distance of geometrical horizon} = \sqrt{2Rh} \qquad (1)$$

(OB $= 2R$ approximately.)

It follows that if the Earth's diameter and the height of the observer's eye above sea level are known, AD, the distance of the observer's geometrical horizon, may be computed.

It will be noticed from Fig. 3 that the dip of the geometrical horizon, denoted by Δ, is equal to the angle at the Earth's centre contained between radii terminating at D and A respectively. We have, therefore:

$$\cos \Delta = \frac{R}{R + h}$$

Since the angle of dip is a small angle:

$$1 - \frac{\Delta^2}{2} = 1 - \frac{h}{R}$$

where Δ is expressed in circular measure; and

$$\Delta = \sqrt{\frac{2h}{R}} \tag{2}$$

The effect of atmospheric refraction is for light coming from the actual horizon—the visible or sea horizon as it is called—to follow a path concave to the Earth's surface as illustrated in Fig. 4.

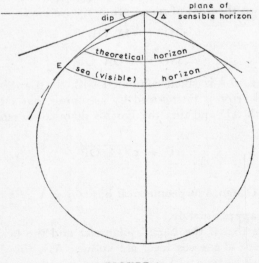

FIGURE 4

Refraction causes the sea horizon to have a greater range than that of the geometrical horizon. It also causes the angle of dip to be smaller than that of the geometrical dip.

The effect of terrestrial refraction on dip and distance of the sea horizon received the attention of many 18th century astronomers and physicists, but there was never general agreement as to the exact effect of refraction. Nevil Maskelyne, under whose direction the first British *Nautical Almanac* was published in 1766 for 1767, stated that one tenth of the theoretical dip should be subtracted from the theoretical dip to give the true dip. Other investigators gave fractions between $\frac{1}{9}$ and $\frac{1}{15}$. At the present time the factor $\frac{1}{13}$ is used, this normally being attributed to the French physicist Biot.

Biot's law is usually expressed thus:

$$r = \frac{d}{13}$$

where r is the angle of terrestrial refraction and d is the distance of the actual sea horizon.

By expressing h in feet and R in nautical miles, and substituting in formulae (1) and (2) we get:

$$\Delta = 1{\cdot}06\sqrt{h}$$

This angle is the theoretical dip in minutes of arc and it is equivalent to the distance of the geometrical horizon in nautical miles.

We shall now investigate the formulae used for finding the actual dip and the distance of the sea horizon using Biot's law, illustrating the investigation by Fig. 5.

In Fig. 5, O represents the observer's eye, and OH lies in the plane of his sensible horizon. Curve OT represents the ray of light, called the *grazing ray*, which enters the observer's eye and which comes from a source on the actual sea horizon, the range of which is equivalent to arc BT, and which is denoted by d. δ is the actual dip. OD and TE are tangents to the grazing ray at O and T which meet at X.

$$XOT = XTO = r = \text{terrestrial refraction}$$

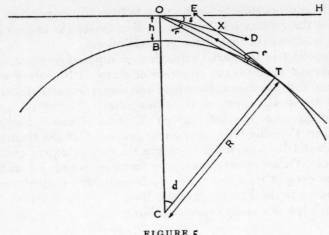

FIGURE 5

Applying the plane sine formula to triangle COT we have:

$$\frac{R}{R + h} = \frac{\sin O}{\sin T}$$

$$= \frac{\sin(180° - d - 90° + r)}{\sin(90° - r)}$$

$$= \frac{\cos(d - r)}{\cos r}$$

$$= \cos d + \sin d \tan r$$

Because d and r are small we may express $\cos d$ as $1 - d^2/2$ and $\sin d$ and $\tan r$ as d and r respectively, without introducing material error. Thus

$$\frac{R}{h + R} = 1 - d^2/2 + rd$$

from which

$$d = \sqrt{\frac{26h}{11R}}$$

Expressing h in feet and R in nautical miles, this formula reduces to:

$$d = 1·15\sqrt{h}$$

Now 1·15 is about $\frac{13}{12}$ of 1·06. It follows, therefore, that refraction extends the distance of an observer's geometrical horizon by about one twelfth of itself.

In the triangle OTC in Fig. 5:

$$C + O + T = 180°$$

therefore,

$$d + (90 - \delta - r) + (90 - r) = 180°$$

from which

$$\delta = d - 2r$$

i.e.

$$\delta = 1 \cdot 15\sqrt{h} - 2 \cdot \frac{d}{13} \quad \text{(Biot's law)}$$

$$= 1 \cdot 15\sqrt{h} - \frac{2}{13} 1 \cdot 15\sqrt{h}$$

$$= \frac{11}{13} 1 \cdot 15\sqrt{h}$$

$$= 0 \cdot 98\sqrt{h}$$

The running of the sea in bad weather causes the sea horizon to be in almost continual vertical motion, and the rising and falling of an observer due to rolling, pitching and heaving of the ship, causes the dip of the sea horizon to be in perpetual change. This trouble may be overcome by taking a series of altitude observations or *shots* and then meaning the results.

The height of eye should be ascertained with precision because an error in dip, which depends upon the observer's height of eye, causes a corresponding error in the altitude. This is of great importance when the observer's eye is near sea level, because the rate of change of dip decreases as the height of eye above sea level increases.

In general, the greater the height of eye the more distinct will be the sea horizon provided that the air is clear. In misty weather, however, when celestial observations may be possible, it is best to observe from a position as near to the sea surface as practicable, so as to bring the sea horizon as near as possible to the observer.

SUN'S SEMI-DIAMETER

The Sun's semi-diameter is the angle at the Earth's centre contained between the true directions of the Sun's centre and

8

circumference or limb. Because the distance of the Sun from the Earth is very great compared with the Earth's radius, the angle at an observer's eye between the true directions of the Sun's centre and his upper or lower limb is not materially different from the angle at the Earth's centre.

When observing the Sun the altitude of his lower limb is measured. It follows, therefore, that an additive semi-diameter correction must be applied in the process of correcting the observed altitude.

The Sun's semi-diameter is greatest when the Earth is at perihelion in early January when its value is 16·3′. It is least in early July when the Earth is at aphelion, when its value is 15·8′. The Sun's semi-diameter is tabulated in the *Nautical Almanac* for 1200 G.M.T. for each day of the year.

MOON'S SEMI-DIAMETER

Because the radius of the Earth is a significant proportion of the distance between the Earth and Moon, the Moon's semi-diameter is sensibly affected by her altitude. The Moon's orbit around the Earth, like that of the Earth's around the Sun, is elliptical, so that the Moon's semi-diameter during any lunation is least when she is at apogee and greatest when she is at perigee. The Moon's semi-diameter is least of all when she is at apogee and has an altitude of 0°. It is greatest when the Moon is at perigee with an altitude of 90°.

Tabulated values of the Moon's semi-diameter given in the *Nautical Almanac* apply to an altitude of 0°, so that a small correction, known as the *augmentation*, is to be applied to the tabulated value.

A formula for finding the augmentation of the Moon's semi-diameter is derived as follows.

In Fig. 6, C represents the Earth's centre and O an observer. M_o represents the Moon on the observer's sensible horizon. M_a represents the Moon whose altitude above the observer's sensible horizon is a. Let the radius of the Earth be denoted by R and that of the Moon by r.

The tabulated value of the Moon's semi-diameter is the angle at the Earth's centre subtended by the Moon's radius. Since M_OO and M_OC are almost equal, the angles θ_O and θ are also almost equal. It follows that the tabulated value of the Moon's semi-diameter, viz. θ, is equivalent to the value when the Moon's centre is on the sensible horizon of an observer.

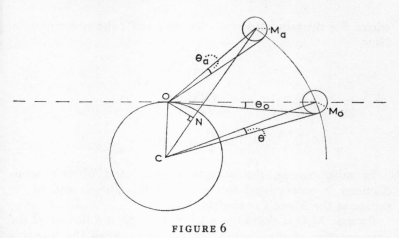

FIGURE 6

Referring to Fig. 6, N is a point on the perpendicular from O on to CM_a.

Now
$$OM_a = NM_a \quad \text{(nearly)}$$
$$= CM_a - CN$$
$$= CM_O - CN$$
$$= OM_O - R \sin a$$

Let the Moon's semi-diameter at altitude a be denoted by θ_a. Then:

$$\theta_a = \frac{r}{OM_a} \text{ radians}$$
$$= \frac{r}{OM_O - R \sin a}$$
$$= \frac{r}{OM_O[1 - (R \sin a/OM_O)]}$$

$$= \frac{r}{\mathrm{OM_O}}\left[1 + \frac{R \sin a}{\mathrm{OM_O}}\right]$$

$$= \frac{r}{\mathrm{OM_O}} + \frac{rR \sin a}{\mathrm{OM_O}^2}$$

$$= \theta + c$$

where θ is the tabulated semi-diameter and c the augmentation. Now

$$c = \frac{rR \sin a}{\mathrm{OM_O}^2}$$

$$= \frac{rR \sin a}{r^2/\theta^2}$$

$$= \frac{R}{r} \sin a . \theta^2$$

In other words, the augmentation of the Moon's semi-diameter is proportional to the sine of the altitude and to the square of the Moon's semi-diameter.

Because M_zO is about one sixtieth of M_zC, it follows that the augmentation of the Moon's semi-diameter when the Moon is in the zenith is about one sixtieth of her semi-diameter. The maximum value of the augmentation of the Moon's semi-diameter is about 0·3′. In practical navigation, therefore, the Moon's augmentation normally is ignored.

PARALLAX

The point on the celestial sphere occupied by a celestial body viewed from a point on the Earth's surface is called the *apparent place* of the body. The point the body would occupy were it viewed from the Earth's centre is called the body's *true place*. The angular distance between the apparent and true places of a body at any instant is called the body's *parallax-in-altitude* at the instant.

Fig. 7 illustrates the Earth whose centre is at C. O is an observer. X is a celestial body.

The parallax-in-altitude of a body such as X is greatest when the body, indicated by X_O, lies on the observer's sensible horizon. This value is called *horizontal parallax* (H.P.). As the altitude of a body increases, its parallax-in-altitude diminishes, so that a body in the zenith, such as X_z in Fig. 7, has zero parallax-in-altitude. This follows because the true and apparent places of a body in the zenith coincide.

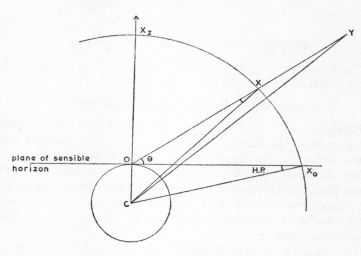

FIGURE 7

It should be clear from Fig. 7 that parallax-in-altitude varies inversely as the distance of the body from the Earth. Parallax-in-altitude for Y in Fig. 7, which has the same apparent place as X, is smaller than it is for X, the nearer body.

The parallax-in-altitude for any celestial body is given by the formula:

$$\text{Parallax-in-altitude} = \text{H.P.} \times \text{cosine altitude}$$

This may be proved with reference to X in Fig. 7 as follows: In Fig. 7 parallax-in-altitude of body X whose altitude is θ, is angle OXC.

Applying the sine formula to triangle OXC we have:

$$\frac{\sin OXC}{OC} = \frac{\sin COX}{CX}$$

i.e.

$$\sin OXC = \frac{OC}{CX} \sin COX$$

$$= \frac{OC}{CX_0} \sin (90 + \theta)$$

i.e.

$$\sin \text{parallax-in-altitude} = \sin \text{H.P.} \cos \theta$$

Since parallax-in-altitude and H.P. are small angles the formula becomes:

$$\text{parallax-in-altitude} = \text{H.P.} \cos \text{altitude}$$

Parallax-in-altitude for a fixed star, on account of its immense distance from the Earth, is minutely small. For practical purposes stellar parallax-in-altitude is ignored. For the Sun, the value of the H.P. varies throughout the year, being greatest when the Earth is at perihelion and least when she is at aphelion. The maximum value of the Sun's H.P. is about 9″. No sensible error results in practical navigation when the Sun's parallax-in-altitude is ignored.

The Moon's parallax-in-altitude is an altitude correction of great importance. The Earth's radius is relatively large compared with the distance of the Moon from the Earth. The ratio between these two distances is about 1/60 so that the Moon's H.P. is about 1°. The Moon's H.P. is greatest when she is at perigee when the value is about 62′, and least when she is at apogee when the value is about 53′.

The oblate shape of the Earth results in the Moon's H.P. being different for different latitudes. The tabulated H.P. is given for an observer located on the equator. For this reason it is called *equatorial H.P.*

In Fig. 8 M_1 represents the Moon on the sensible horizon of an observer o_q on the equator. The angle $o_q M_1 C$ is the equatorial H.P. M_2 represents the Moon on the sensible horizon of an observer o_p at the Earth's pole. The angle $o_p M_2 C$ is the *polar H.P.*

In any latitude other than 0°, the tabulated H.P. must be reduced by an amount called the *reduction to the Moon's H.P.*

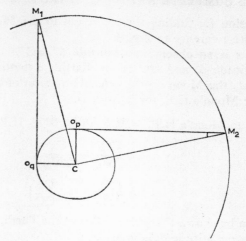

FIGURE 8

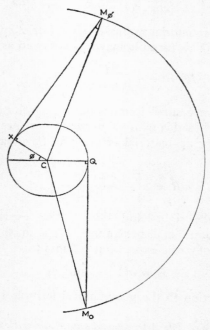

FIGURE 9

An expression for finding the reduction to the Moon's H.P. is derived with reference to Fig. 9.

In Fig. 9 x is an observer in latitude $\phi°$. xC is the Earth's radius in latitude ϕ, and QC is the Earth's equatorial radius. $M\phi$ represents the Moon on the sensible horizon of observer x. $xM\phi C$ is the Moon's H.P. for latitude ϕ.

Reduction to Moon's H.P. = H.P. in lat $0°$ − H.P. in lat ϕ

$$= \frac{CQ}{CM_O} - \frac{Cx}{CM_\phi}$$

$$= \frac{CQ}{CM_O} \left(1 - \frac{Cx}{CQ}\right)$$

Now it can be shown that† the radius of the Earth, R, for any given geocentric latitude θ, is given by:

$$R = \frac{a(1 - c)}{[\cos^2 \theta(1 - c)^2 + \sin^2 \theta]^{1/2}}$$

where a is the equatorial radius of a terrestrial spheroid having an ellipticity c. This formula may be expressed as:

$$R^2 = \frac{a^2(1 - c)^2}{\cos^2 \theta(1 - c)^2 + \sin^2 \theta}$$

Now c is a very small fraction having a value of about $\frac{1}{300}$. The terms in c^2 in this expression may, therefore, be neglected without introducing material error. The formula thus reduces to:

$$R = a\left[\frac{1 - 2c}{1 - 2c \cos^2 \theta}\right]^{1/2}$$

Expanding the right-hand side of this expression by the Binomial Theorem and neglecting terms in the second and higher powers of c, the expression reduces to:

$$R = a(1 - c \sin^2 \theta)$$

Now the reduction to the geographical latitude is a very small

† See pp. 49–50 of *The Astronomical and Mathematical Foundations of Geography* by C. H. Cotter.

angle and it may be assumed that the geographical latitude ϕ is equivalent to the geocentric latitude θ so that:

$$R = a(1 - c \sin^2 \phi)$$

or,

$$\frac{R}{a} = \frac{Cx}{CQ} = (1 - \sin^2 \phi/300)$$

It follows that:

Reduction to Moon's H.P. = Equatorial H.P. $\times \sin^2 \phi/300$

The reduction to the Moon's H.P. is greatest for latitude 90°, but it never exceeds about 0·2′. It is, accordingly, ignored by practical nautical astronomers except when navigational refinement is sought.

It is interesting and important to note that although the Moon's H.P. and semi-diameter are constantly changing, their values are always in constant ratio. This fact, which facilitates the construction of Moon altitude correction tables, is proved with reference to Fig. 10.

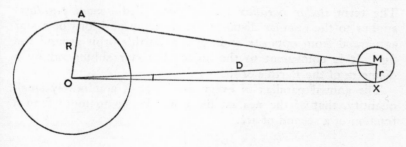

FIGURE IO

In Fig. 10, O and M represent the centres of the Earth and Moon respectively. Let the radii of these bodies be denoted by R and r respectively. Let the distance between their centres be denoted by D.

Now   AMO = Moon's H.P.

and   MOX = Moon's S.D.

These are small angles, therefore:

$$\frac{R}{D} = \text{Moon's H.P.}$$

and

$$\frac{r}{D} = \text{Moon's S.D.}$$

from which:

$$D = \frac{r}{\text{Moon's S.D.}} = \frac{R}{\text{Moon's H.P.}}$$

Therefore:

$$\frac{\text{Moon's S.D.}}{\text{Moon's H.P.}} = \frac{r}{R}$$

The ratio between r and R is constant and it is about 1/4, therefore the ratio between the Moon's semi-diameter and her horizontal parallax is also constant.

STELLAR PARALLAX

The term *stellar parallax* (sometimes called *annual parallax*) applies to the angular distance between the positions of a star as viewed from opposite ends of the Earth's orbit around the Sun. It is equivalent to the angle at a star subtended by a diameter of the Earth's orbit.

The annual parallax of every fixed star is a minutely small quantity, that of the nearest fixed star being no more than a fraction of a second of arc.

ABERRATION, PRECESSION AND NUTATION

The 17th- and early 18th-century attempts made by philosophers, including Jean Picard, Robert Hooke and James Bradley, to discover the annual parallax of a fixed star, led to the discovery of an astronomical phenomenon known as *aberration of light*. This phenomenon is due to the Earth's orbital speed being a relatively big proportion of the speed of light. The orbital motion of the Earth is sufficiently fast to cause the light from a star to shift slightly in the direction in which the Earth is moving. Because of this a fixed star, during the course of a year, appears to describe an ellipse the centre of which is the true place of the star.

The effect of aberration on a star's position may be as much as to cause a displacement of about 20″ of the star's true place. By noting the changes that take place during the year, in the declination and Sidereal Hour Angle of some fixed stars, the effect of aberration can be detected from the *Nautical Almanac.*

In addition to aberration a star's celestial position may be affected by the *real motion* of the star. That component of a star's real motion across an observer's line of sight is called the star's *proper motion.* Precession of the equinoxes and nutation may also affect a star's celestial position.

Because the Earth is a spinning body she possesses the property known as *gyroscopic inertia.* This is the expression of the tendency a spinning body has to maintain its plane of spin. Every spinning body maintains its plane of spin so long as the body is not influenced by an external couple acting upon it. An external couple acting upon a spinning body causes the axis of spin of the body to trace out a conical movement the period of which is usually very long compared with that of the rotation of the body. This motion is called *precession.*

The revolution of the Earth around the Sun, the force of attraction between the Earth and the Sun, the Earth's oblate shape, and the fact that the plane of the Earth's spin is inclined to that of her orbit around the Sun, results in the Earth's axis precessing, the period of precession being about 26,000 years.

The celestial poles, because of the precession of the Earth's axis, would trace out small circles centred at the poles of the ecliptic, each having a spherical radius of about $23\frac{1}{2}°$. The precessional movement has a retrograde direction which results in the gradual increase in the Right Ascension of all fixed celestial bodies.

The Moon has a similar effect to that of the Sun known as *nutation.* Nutation results in the celestial poles describing wavy circles around the poles of ecliptic, each wave being completed in a period of about $18\frac{1}{2}$ years.

As a result of precession and nutation the celestial positions (R.A. and declination) of all fixed celestial positions change with the passage of time.

IRRADIATION

When a bright object is viewed against a darker background, the bright object appears to be larger than it actually is. On the other hand, an object viewed against a lighter background appears smaller than it actually is. This optical phenomenon is known as *irradiation*.

The celestial bodies, when viewed against the relatively darker sky, are affected by irradiation. Moreover, the sea horizon, because the sky is generally brighter than the sea, appears to be depressed on account of irradiation.

When the Sun's upper limb is observed the effect of irradiation is apparently to lower the sea horizon and to elevate the Sun's limb. The combined effects make it necessary to apply an irradiation correction amounting to $-1 \cdot 2'$ to altitude observations of the Sun's upper limb.

When the Sun's lower limb is observed, the two effects of irradiation tend to neutralize one another so that no irradiation correction is considered to be necessary.

PERSONAL ERROR

The timing of an event such as the instant when the image of a celestial body makes contact with the horizon, is affected by the temperament and nervous and physical condition of the observer. Any error due to this cause is called *personal error* or *personal equation*.

Personal error may be detected by comparing observations with those made by other observers. Personal error varies not only between observers, but it may vary at different times for any one observer.

CHAPTER III

The Astronomical Position Line

We have, in Chapter 1, defined an astronomical position line as the projection on a navigational chart or plotting sheet of part of a circle of equal altitude somewhere on which the navigator may fix his ship's position.

A *circle of equal altitude* is a circle on the Earth's surface centred at the G.P. of an observed celestial body: and the G.P. is the point on the Earth at which the body is in the zenith. At the zenith of the G.P. lies the observed object. This, in turn, occupies one of the corners of the astronomical triangle. It is convenient to 'bring the astronomical triangle down to Earth' in order to facilitate relating it to the astronomical position line which is obtained after solving the PZX triangle.

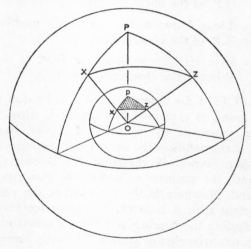

FIGURE I
115

Fig. 1 represents the celestial sphere with the Earth at its centre. p is the Earth's North Pole and P is the elevated celestial pole. X is an observed celestial body whose geographical position is at x. z represents the position of an observer (which position is generally unknown to him). O is the Earth's centre, and PO, ZO and XO are straight lines of projection by means of which the PZX triangle is projected on to the Earth's surface. The projected pzx triangle is geometrically similar to the PZX triangle. In the celestial triangle PZX:

PZ = co-altitude of celestial pole

PX = polar distance of the observed object

ZX = zenith distance of the observed object

angle P = local hour angle of the observed object

angle Z = azimuth of the observed object

In the terrestrial triangle pzx:

pz = co-latitude of the observer

px = co-latitude of the G.P. of the observed object

zx = great circle distance between the observer and the G.P. of the observed object

angle p = d.long between the observer's position and the G.P. of the observed object

angle z = great circle bearing of the G.P. of the observed object at the observer's position.

If the G.M.T. of the observation of X is known, the G.P. of X may be found as explained in Chapter I. Also, the arc ZX having been measured by means of a sextant (see Part III, Chapter I), corresponds to the great-circle distance zx; and this is, accordingly, known. If, therefore, the bearing of z from x can be found, the position of x, the observer's position, can also be found. Nowhere is the meaning of the small word *if* more important than it is in the foregoing sentence. There is no way of finding the bearing of z from x unless at least three parts of the pzx triangle are known. The only known parts are the sides px, which is equal to PX the polar distance of the observed object, which can be found from the *Nautical Almanac*; and zx, which is equal to ZX. This, as stated above, is determined by sextant measurement. The third side of the pzx triangle is not known unless the observer knows his latitude. Because the observer is at sea he does not generally know this at the time of an observation. The angle at z, which is equal to PZX, cannot be found from a compass observation to a degree of accuracy commensurate with that required for finding the ship's position. The angle at p, which is equal to the angle ZPX the local hour angle of the observed object, cannot be found unless the observer's longitude is known. And again, because the observer is at sea he does not generally know his longitude at the time of the observation.

If the navigator *does* know his ship's latitude precisely it is an easy matter for him to find his ship's longitude from an altitude observation. If the ship's latitude is known, the three sides of the astronomical triangle are also known from a single observation, so that by the rules of spherical trigonometry, the angle P may be computed. This angle, being the local hour angle of the observed body, when compared with the G.H.A. of the body for the time of the observation, will give the ship's longitude.

On the other hand, if the navigator knows his ship's longitude, it is an easy matter for him to find his ship's latitude from an altitude observation. In this case, the angle P in the astronomical triangle is known in addition to the sides PX and ZX, so that from the rules of spherical trigonometry it is possible to compute the side PZ of the astronomical triangle from which the observer may find the ship's latitude.

There are special circumstances whereby a navigator may ascertain his ship's latitude without knowing her longitude or the ship's longitude without knowing her latitude. Latitude may be found from an observation of a celestial body on the celestial meridian of the observer, in which case the observer's longitude need not be considered. Longitude may be found from an observation of a celestial body on the observer's prime vertical circle (this being the vertical circle passing through the east and west points of the observer's horizon), in which case the observer's latitude need not be considered. These particular problems, however, do not invalidate the general rule in nautical astronomy that an observer may find *either* latitude *or* longitude, but not both latitude and longitude from a single observation.

ASTRONOMICAL POSITION CIRCLES

The zenith distance of an observed celestial body in minutes of arc is equivalent to the radius in miles of the circle of equal altitude centred at the G.P. of the body. If the G.M.T. of an observation is known and a *Nautical Almanac* is available, the G.P. of the observed object may be found.

If simultaneous observations of the altitudes of two celestial objects are made and the G.M.T. of the simultaneous observations noted, sufficient information is available to the observer for him to find his ship's position.

A simple method of fixing a ship from simultaneous time altitude observations is to plot the G.P's of the two observed objects on a model globe; and then to draw the projections of the two circles of equal altitude to intersect at the ship's position. This position is then lifted from the model globe using the graticule of lines of latitude and longitude. This method, although attractive and simple in theory, is not practicable because of the difficulty of obtaining a position to the necessary degree of accuracy.

Because the Mercator chart virtually replaces the model globe, it is natural to inquire into the nature of the projections of circles of equal altitude on the Mercator chart. Investigation in this direction reveals that, in general, the projected circle of equal altitude is a complex curve the form of which depends upon the relative values of the altitude and declination of the observed body. The closed form of the projection of an equal-altitude circle on a Mercator chart resembles an ellipse. The smaller is this ellipse the more closely it resembles a circle. In this case the projection of the circle of equal altitude may be considered to blend with a circle when the observer's latitude and the zenith distance of the observed object are both small, and it is an easy matter to plot such a circle of equal altitude on a Mercator chart. All that has to be done is to plot the G.P. of the observed body for the time of its observation and, with this point as centre, describe a circle of radius equa[l] miles to the zenith distance of the observed body in min[utes] of arc.

This method of drawing what may be described as a position circle is the only practical result afforded by the study of the forms of the projection of a circle of equal altitude on a Mercator chart. But, as pointed out by the French astronomical navigators of the 1870's after an intense and significant study of the seaman's nautical astronomical problems, it is a precious result because it permits of the very simple and rapid utilization of altitude observations of celestial bodies at great altitude.

It is not uncommon in cloudy weather for a star to appear near the zenith when it is impossible to see others in a less elevated position. Moreover, within the tropics, the Sun often attains a very high altitude near his time of meridian passage. On occasions an observation in these circumstances may afford a navigator the means of fixing his ship when, perhaps, other methods of fixing are not available.

The method of fixing by plotting astronomical position circles on a large-scale Mercator chart was first suggested in about 1874 by the French naval officer Aved de Magnac who, with other French navigators and astronomers of the time, played a significant role in the advancement of nautical astronomy. The attention of British navigators was directed to the problem by Captain T. S. Angus of the P. & O. Company in the year 1884, and for many years the method was known as 'Captain Angus's method'.

Angus's method involves observing two altitudes (and corresponding G.M.T's) of the Sun when he is high in the sky near the time of his meridian passage. The interval of time between the observations must be sufficiently small for the position circles corresponding to each of the two observations to intersect, thus enabling the observer to find his ship's position by plotting. If the interval between the times of the observations is more than a few minutes, it might be necessary to transfer the first position circle, and to treat the problem as a running fix.

The following example serves to illustrate this simple and effective method of fixing the ship when astronomical conditions permit.

9

EXAMPLE: The following information was obtained from observations of the Sun near meridian passage before noon on August 7th 1968.

Observation No. 1: True altitude = 89° 12'

G.M.T.   = 16 hr 20 min 15 sec

Observation No. 2: True altitude = 89° 22'

G.M.T.   = 16 hr 23 min 11 sec

Find the ship's position at the time of the second observation. From the *Nautical Almanac*:

Sun's declination = 16° 17·4' N.

Observation No. 1: G.H.A.T.S. = 63° 39'

Observation No. 2: G.H.A.T.S. = 64° 23'

From this information the Sun's G.P. at each observation may be plotted. Fig. 2 illustrates part of a Mercator chart on which the navigator plots the position circles in order to fix his ship.

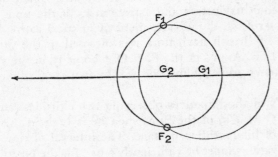

FIGURE 2

In Fig. 2:

G_1 represents the G.P. of the Sun at the time of the first observation. This position is:

lat 16° 17·4' N. = declination of the Sun

long 63° 39' W. = G.H.A. of the Sun at first observation

G_2 represents the G.P. of the Sun at the time of the second observation. This position is:

lat 16° 17·4′ N. = declination of the Sun

long 64° 23′ W. = G.H.A. of the Sun at second observation

The radius in miles of the position circle centred at G_1 is equal to the zenith distance of the Sun at the time of the first observation in minutes of arc. This is 48 miles, i.e. (90 − 89° 12′).

The radius of the position circle centred at G_2 is 38 miles, i.e. (90 − 89° 22′).

The d.long G_1G_2 is equivalent to the difference between the G.H.A's of the Sun at the times of the two observations. This is equivalent to the difference between the G.M.T's of the observations. This is 2 min 56 sec or 44 minutes of arc.

Because the interval between the times of the observations is small, the observations are treated as being simultaneous, so that the first position circle has not been transferred.

The two position circles intersect at F_1 and F_2 one of which represents the ship's position at the time of the second observation. The navigator would be able to decide which of the two points of intersection is the fix from knowledge of his ship's D.R. position or, failing this, by noting the bearings of the Sun at the times of observation. Had the bearings been southerly, the observer would have been at F_1. Had they been northerly the fix would have been at F_2.

Captain Angus's method is severely limited in its application. In the general problem of nautical astronomy the radius of the circle of equal altitude is very big, usually in the order of many hundreds or even thousands of miles, so that the position circle cannot readily be projected on a Mercator chart.

The solution to the general problem of nautical astronomy involves finding a point through which to draw an astronomical position line. The required position line, the direction of which is at right angles to the bearing of the observed object, is a small fragment of a position circle. This being so no material error is introduced by assuming it to be a straight line on a Mercator chart.

There are two general methods of obtaining an astronomical

position line. One method stems from a discovery made by the American Captain Thomas Sumner in 1837. The other developed from the excellent investigations, referred to above, made in France during the last century, and to which the name of Marcq Saint Hilaire is closely associated. We shall deal with each of these methods historically.

SUMNER'S METHOD AND ITS MODIFICATION

The history of what has become known as *astronomical position-line navigation* is full of interest. It is usually regarded as having begun in 1837 with Captain Sumner's discovery. There seems to be no doubt, however, that for some decades before this date scientific navigators of many nations had given considerable attention to the problems of nautical astronomy. Position-line, or intersectional, navigation, as it used to be called, grew from the method of finding latitude from two astronomical observations—a method known as the double altitude. The germ of astronomical position-line navigation is to be found in a work on navigation by Samuel Dunn published about 1780, in which the author introduced a problem entitled:

'Of a general method whereby the latitude may be found having any two altitudes of the Sun and the time elapsed between the observations'.

By assuming two latitudes differing about a degree or less, and not widely different from the latitude by D.R., Dunn showed that the two altitudes give four hour angles, two of which pertain to each of the assumed latitudes. He then made the following statement:

'As the difference of the elapsed times computed from the assumed latitudes is to the difference of those latitudes: so is the difference between the true elapsed time to a number of minutes which, added to or subtracted from the corresponding assumed latitude, as the case requires, gives the true latitude required when the latitudes are assumed near enough for the truth.'

Chronometers were scarce in the days when Dunn introduced

this novel problem. Had they been common it is likely that Dunn would have extended his method for finding longitude as well as latitude.

Dunn's resolution of the double-altitude problem had been discussed in all its aspects by the French astronomer Lalande, but there is much justification for believing that the development of position-line navigation from Dunn's time onwards, was carried out on the basis of his double-altitude problem.

In 1833, Commander Thomas Lynn of the East India Company published a method for finding latitude *and time* by double altitudes based on Dunn's method. Lynn's method, similar to one used by officers in the British Royal Navy at the beginning of the 19th century, was known as the *method by trial and error*.

The discovery of position-line navigation rightly belongs to Sumner, who is credited with being the first to systematize the problem of finding position at sea from astronomical observations.

Sumner's discovery was made in 1837 during a voyage from Charleston to Greenock. The details of his discovery are given in a pamphlet first published in Boston in 1843.

Sumner pointed out that when knowledge of the latitude is uncertain there are only two instants during the day at which the Sun's altitude can be used to find the longitude if the G.M.T. is known; and that there is only one instant each day when the Sun's altitude can be used for the latitude, unless the local hour angle of the Sun is accurately known. At all times when the Sun is not at meridian passage or on the prime vertical circle errors of latitude and longitude proportional to the angular distance of the Sun from north or south, and east or west, respectively may be great. He then described how a single altitude of the Sun, taken at any time, may be used to determine a line diagonal to, and affording a substitute for, a parallel of latitude or a meridian. This line, when plotted on the chart, is the astronomical position line we have discussed.

In preposition-line navigation days, the customary method of finding position at sea was to observe the Sun on the prime vertical (or as near to the prime vertical as latitude and declination permitted) to find the longitude at morning sights corresponding to a D.R. latitude, and then to observe the Sun on the meridian to ascertain the noon latitude. The morning longitude was run up to noon from knowledge of the course and distance made good between the times of morning sights and noon to find the longitude at noon.

The latitude used in solving the longitude from a Sun observation was generally different from the ship's actual but unknown latitude. In some cases the difference or error in latitude was ascertained from the noonday Sun observation, in which case the morning sight was reworked using the correct latitude. In the event of the sky being overcast at noon, the morning sights for longitude were often discarded as useless.

Sumner discovered that a single observation of the Sun (or other celestial body), even if the latitude is uncertain, is of value. He demonstrated that a single altitude taken at any time is sufficient to obtain an astronomical position line.

It seems, from what Sumner wrote, that his discovery was something of an accident. He relates how, after passing the meridian of 21° W. when eastbound across the north Atlantic, no astronomical observations were obtained until near the land:

'On the 17th December 1837', he wrote, 'the ship was kept on ENE. under short sail with gales. At about 10 a.m. an altitude of the Sun was observed and the chronometer time noted; but, having run so far without any observation, it was plain that the latitude by D.R. was liable to error, and could not be entirely relied upon. Using however this latitude in finding the longitude by chronometer, it was found to put the ship 15′ of longitude East from her position by D.R. which in latitude 52° is 9 nautical miles. This seemed to agree tolerably well with the D.R.; but feeling doubtful of the latitude the observation was tried with a latitude 10′ further north. Finding this placed the ship ENE. 27 miles of the former position, it was tried again with a latitude 20′ north of the D.R. This also placed the ship still further ENE. and still 27 miles. These three positions were then seen to lie in the direction of Small's Light. It then at once appeared that the observed altitudes must have happened at all the three points and at the Small's Light, and at the ship, all at the same time; and it followed that Small's Light must bear ENE. if the chronometer was right. Having been convinced of this truth, the ship was kept on her course ENE., and in less than an hour Small's Light was made.'

The first astronomical position line may, therefore, be regarded as Sumner's line of position through the Small's Lighthouse. It is interesting to note that this first astronomical position line was used to fetch up a known position on the land— a very valuable use of a single position line.

The principles of Sumner's method is illustrated in Fig. 3.

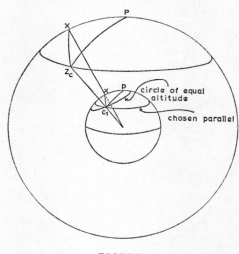

FIGURE 3

Fig. 3 illustrates the celestial sphere with the Earth at its centre. p is the Earth's North Pole and P is the elevated celestial pole. X is an observed celestial body and x is its geographical position. Point c_1 whose zenith is at Z_c, lies on both the circle of equal altitude and a parallel of latitude near to the ship's actual, but unknown, parallel of latitude.

In the astronomical triangle PZ_cX:

> PX = polar distance of observed object
>
> PZ_c = (90° − latitude of chosen parallel)
>
> Z_cX = zenith distance of the observed object

These three sides of the triangle are known, so that the angle P may be computed. This will give the hour angle of X at the meridian of c. Comparing this with the G.H.A. of X, the longitude of c may be found.

The problem is then reworked for position c_2 which, as was the case with c_1, lies on the circle of equal altitude but on a different parallel of latitude from that on which c_1 lies. Fig. 4 illustrates an enlargement of the Earth as it appears in Fig. 3 and shows, in addition to c_1, the second point c_2.

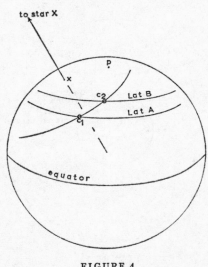

FIGURE 4

The two positions c_1 and c_2 are plotted on a large-scale navigational chart, and a straight line drawn through them. This straight line is regarded as being the required astronomical position line. Two such lines, provided that they intersect, enable a navigator to fix his ship. Fig. 5 illustrates two intersecting circles of equal altitude obtained from simultaneous observations of two heavenly bodies.

The points c_1 and c_2 in Fig. 5 are those ascertained from the observation of the star X. The points c_3 and c_4 are those ascertained from a simultaneous observation of a star Y. The points

c_1 and c_3 are on the parallel of latitude A. Points c_2 and c_4 lie on parallel of latitude B. Latitudes A and B are chosen in relation to the ship's D.R. latitude. The normal practice was to choose two latitudes which embrace the ship's D.R. latitude, one 10' to the north and the other 10' to the south.

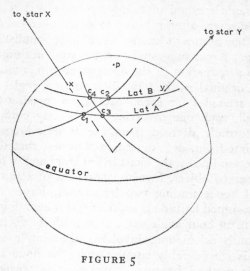

FIGURE 5

Fig. 6 illustrates the manner of plotting the position lines obtained from the observations to which Fig. 5 applies.

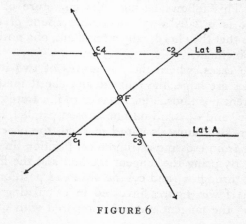

FIGURE 6

Referring to Fig. 6, the ship's position lies at the point F at the intersection of the two position lines.

The straight line joining c_1 and c_2 (or c_3 and c_4) does not, theoretically, coincide with the circle of position. It is a straight line joining two points on the position circle. It is for this reason that Sumner's method of fixing by crossing two astronomical position lines was often referred to as the *chord method*.

Soon after Sumner's method had been published the illustrious Henry Raper suggested a modification of Sumner's method leading to a reduction in the amount of computation required. Sumner's original method involves solving angle P in each of two PZX triangles for each observation, making four calculations in all, each one involving a relatively complex trigonometrical formula derived from the spherical cosine formula. Raper pointed out, as indeed had Sumner, that the projection of the position line on the chart lies at right angles to the bearing of the observed object. Instead, therefore, of assuming two latitudes and hence finding two longitudes, Raper demonstrated that one assumed latitude is sufficient, this enabling the observer to calculate an hour angle and an azimuth for each observation. The azimuth calculation, involving the simple spherical sine formula, is considerably simpler than the hour angle calculation. This method became known as the *tangent method* in contrast to Sumner's original 'chord method'.

In practice both the chord and tangent methods give the same result. This follows because the curvature of circles of equal altitude is usually very small on account of their radii being large so that the chord, tangent and arc, of a position circle are almost coincident.

In extreme cases, where the curvatures of two intersecting position circles are large, the tangent and chord methods yield slightly different positions. Fig. 7 serves to illustrate this.

In Fig. 7, c_1 and c_2 stand on one chosen parallel, and c_3 and c_4 on another. By Sumner's chord method, the observed position resulting from crossing astronomical position lines c_1c_4 and c_2c_3 is at F_c. By using the tangent method and the first chosen latitude (that through c_1 and c_2) the observed position is at F_t.

Despite the fewer figures involved in calculating the ship's position using the tangent method compared with the number used in the chord method, the latter method appears to have been the more popular amongst the generality of seamen until the beginning of this century. The increasing popularity of the Azimuth tables designed by Burdwood and Davis, and the *A, B and C Tables*, which are still very popular amongst merchant seamen, spelt doom to the old chord method; and when the B.O.T. examiner ceased to ask questions about it (at a time long after it had outlived its usefulness), it suffered a natural death.

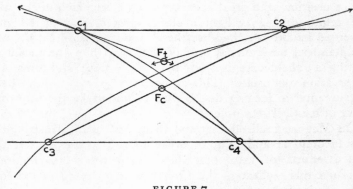

FIGURE 7

Despite its inferiority as a method for fixing a ship, compared with the alternative general method of finding a position line, the tangent method still maintains its popularity amongst merchant seamen. Perhaps this is so because they are taught the method to satisfy the examiner; and, having been taught it, and finding it suits their needs, they persist in using it to the exclusion of other methods.

MARCQ SAINT HILAIRE'S METHOD

In 1875, a paper in a French scientific journal entitled *La Nouvelle Navigation* appeared under the name of a French naval officer named Marcq St. Hilaire. He was the inventor of a navigational method known to the French as *Méthode du Point Rapproché*, and to British navigators as the *Intercept Method*.
The work of the French naval officers and astronomers who

introduced the methods known as the New Navigation, marked a collective, and very fruitful, attempt to study the problems of astronomical navigation in a scientific way.

Commander Aved de Magnac, while serving as navigator in a French man-o'-war about 1867, found himself in circumstances not dissimilar to those in which Captain Sumner found himself on that eventful day in 1837 when astronomical position-line navigation first saw the light of day. Circumstances were such that it was very important for de Magnac to find his ship's position with great accuracy. In a clear patch of sky near the zenith during twilight a star appeared, and de Magnac observed its altitude. Because the altitude was so great none of the standard navigational methods afforded him the means of getting a reliable result from his observation; and thus, the observation was useless. This observation appears to have been instrumental in leading de Magnac to undertake the improvement of nautical astronomy.

De Magnac realized that the theory of nautical astronomy was incomplete and that, in order to rectify the matter, the united efforts of astronomer and navigator were required. Through this realization the French were to stand in the fore of navigational improvements for many decades.

On returning to France de Magnac, with the co-operation of Villarceau, a prominent French astronomer, published an important book on navigation entitled *Nouvelle Navigation Astronomique*. It was in this book that the navigator's problem was first properly defined. The authors pointed out that the nautical astronomical problem *par excellence* is the determination of a POINT. Hitherto from the time of Sumner's discovery all investigations into the nautical astronomical problem had been related to the determination of a LINE; and, clearly, the problems related to point and line are different. Here then was an entirely fresh approach into nautical astronomy.

In the first place, de Magnac and Villarceau investigated the theory of the single observation, and showed incontestably that fixing a ship by astronomical methods requires more than one observation. Secondly, they showed that computation is necessary, projection of circles of equal altitude on the chart not providing a practical solution to the problem. They then derived formulae which, although of no practical use to the navigator, were instrumental in demonstrating the most favourable conditions for finding a ship's position by astronomical methods.

They showed, as others had shown before them, that for best results:

1. The difference of azimuths of the observed objects is 90°.
2. The altitudes must not be too great.

The first condition follows from an investigation into the errors of altitude and their effects on the resulting fix (see Chapter VI). The second condition arises from the fact that if the altitude is great, the circle of equal altitude is small, in which case the points of intersection of the two circles of equal altitude are close together and the mathematical solution may be indeterminate.

Further investigation led to the *Méthode du Point Rapproché*, credit for the invention of which is given to Captain (later Admiral) Marcq St. Hilaire, an officer endowed with a profoundly acute mathematical mind.

The intercept method is similar to the tangent method which we have discussed above. Fig. 8 serves to show the underlying principle of the method.

The point E, in Fig. 8, represents the ship's D.R. position. Let the circle of radius r be a circle of error within which it is assumed that the actual position of the ship lies. AA_1 represents part of a circle of equal altitude which cuts the circle of error at Y and Y_1. The ship's actual position must lie on the arc YXY_1 provided that the altitude of the body has been correctly observed. Although the exact position of the ship on arc YXY_1 is unknown; the point X, which lies midway between Y and Y_1 in a direction from E corresponding to the azimuth of the observed body (or 180° away from it had the curve AA_1 been concave instead of convex, to the right), is the most likely position of the ship. Hence the name *point rapproché* given to the point X.

The *point rapproché* lies at a distance from E called by British navigators the *intercept*, and the name given to the point is *intercept terminal position*. The intercept terminal position is coincident with E when the D.R. position happens to coincide

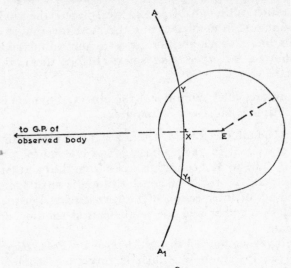

FIGURE 8

with the ship's actual position, in which case the intercept is zero. In other circumstances the *point rapproché* must be nearer to or farther from the geographical position of the observed body than is the point E. In Fig. 8 X is nearer to the G.P. of the observed body than is E. In this case the intercept is named TOWARDS. Had the circle of equal altitude been concave instead of convex to the right in Fig. 8, the point X would have been farther from the G.P. of the observed body than E, in which case the intercept would have been named AWAY.

It is clear from Fig. 8 that if the navigator has a single observation he is able to find the position of the *point rapproché*. If he takes this as an approximate position of his ship instead of taking the D.R. position, he makes use of a position nearer to the ship's true position than is the D.R. position.

Fig. 9 illustrates how the *point rapproché* may be found.

Fig. 9 illustrates the Earth. P is the Earth's North Pole and E is the D.R. position of an observer who observes a celestial object the G.P. of which is at S. The arc of the circle of equal altitude YXY$_1$ is that on which the observer assumes he must lie—the small circle representing the circle of maximum error.

Knowing the latitude of E the arc PE is known, this being the co-latitude of E. Knowing the longitudes of E and the G.P. of the observed body, the angle P, which is the local hour angle of the observed body at E, may be found. The side PS of the spherical triangle PSE is known, this being equivalent to the polar distance of the observed body. With sides PS and PE and the included angle P, the side SE may be calculated using the spherical cosine formula or a formula derived from it. SE is a measure of the great-circle distance between the D.R.

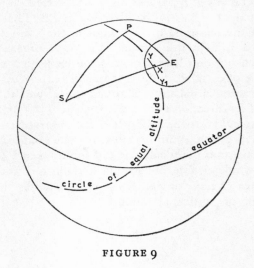

FIGURE 9

position E and the G.P. of the observed body. This is equivalent to the zenith distance of the body at the D.R. position, and is referred to as the Calculated Zenith Distance. The arc SX is the radius of the circle of equal altitude. It is, accordingly, equal to the zenith distance of the body at the observer's position at the time of the observation. This is equivalent to the great-circle distance between the *point rapproché* X and the G.P. of the observed body S, and is called the Observed Zenith Distance.

The difference between arcs ES (which is equivalent to the Calculated Zenith Distance or C.Z.D.), and XS (which is equivalent to the Observed Zenith Distance or O.Z.D.), is equal to the intercept, which is arc EX.

$$\text{Intercept} = \text{C.Z.D.} \sim \text{O.Z.D.}$$

If the C.Z.D. is greater than the O.Z.D. the intercept is named TOWARDS. If the C.Z.D. is less than the O.Z.D. the intercept is named AWAY.

By plotting the D.R. position on the chart and drawing a line of correct length and in the correct direction to represent the intercept, the *point rapproché* may be plotted. The trend of YXY_1 is at right angles to the direction of ES. Therefore, if the azimuth of the observed body is known, the arc YXY_1 may be plotted as a straight line through the *point rapproché*. This is the required position line, which is the projection of a tangent to the circle of equal altitude at the *point rapproché*.

The intercept is usually a short distance, and the degree of accuracy of its direction need only be coarse. It is not necessary, therefore, to *calculate* the azimuth of the observed body in order to find this direction. Azimuth tables may be used instead. These tables contain an orderly collection of solutions of PZX triangles for every whole degree of latitude and declination and every four minutes of hour angle. Their use facilitates the position-line problem of nautical astronomy.

CHAPTER IV

The Latitude

In the days before chronometers and nautical almanacs, the observation of the Sun on the celestial meridian ranked as the most important of all astronomical observations. With the advent of the chronometer, perfected by the Yorkshire carpenter John Harrison in the mid-18th century, the way was open for the navigator to find longitude, as well as latitude, at sea. With the introduction of the British *Nautical Almanac*, which the Astronomer Royal Nevil Maskelyne published in 1766, the seaman was provided with astronomical data presented in a way that made it relatively easy for him to find longitude from a lunar observation. It is interesting to note that the chronometer and the *Nautical Almanac* made their appearances almost simultaneously.

Chronometers for a long time after they became available were expensive and, therefore, scarce. The lunar problem, for about a century after its introduction, appears to have been the standard method for finding longitude at sea by astronomical methods. The problem of the longitude, which the early nautical almanacs were specifically designed to facilitate, involved measuring, by means of a sextant, the angle between the Moon and the Sun or a selected star lying in or near the Moon's monthly circuit of the heavens. The observed lunar distance had then to be *reduced*, by which is meant computing the angle at the Earth's centre between straight lines terminating respectively at the Moon's centre and that of the second body. The process of doing this, in which corrections for refraction and parallax were to be made, was called *clearing the lunar distance*. The cleared lunar distance was then to be compared with tabulated distances given against G.M.T. in the *Nautical Almanac*. Having found the G.M.T., an ordinary altitude sight, taken at the time at which the lunar distance was measured, enabled the observer to find his ship's longitude provided that the ship's latitude was known.

The lunar distance problem was not at all an easy problem for most navigators. So that until the time when chronometers did become common—about the middle of the 19th century—and the methods of modern position-line navigation had been discovered, the generality of navigators observed the noonday Sun for latitude, and relied largely on D.R. navigation for longitude.

The noonday Sun observation has lost some of its former glory since the advent of position-line navigation. Now that an astronomical position line may be ascertained at any time provided that an altitude observation of any navigational celestial body is possible, there is no reason whatever to rely solely on the meridian altitude observation of the Sun at noon. However, the ease with which the latitude at noon may be found from a meridian altitude observation of the Sun, coupled with the fact that the organization of clock time on many ships is related to apparent time for the noon meridian, has resulted in the Sun maintaining its rank, in the eyes of many of the more conservative seamen of our time, as the pre-eminent celestial body for the purpose of astronomical navigation.

The observation of the altitude of any navigational celestial body when that body is at meridian passage, affords an easy method of finding the observer's latitude. Such an observation is known as a *meridian altitude observation.*

We have seen in Part I, Chapter V, that any celestial body, in performing its apparent diurnal motion, attains its greatest altitude when it bears due north or south. When a body bears due north or south it lies on the observer's celestial meridian, at which time it is said to *culminate* or *transit*, or to be at *meridian passage.*

In the case of an altitude observation of a celestial body bearing north or south, the two sides PZ and PX of the astronomical triangle coincide. In other words, the hour angle of a body at meridian passage, being 00 hr 00 min 00 sec or 0° 00', results in the collapse of the PZX triangle into an arc of a great circle.

Because the direction of an astronomical position line is at right angles to the azimuth of a body at the time of observation, the position line obtained from an observation of a celestial body at meridian passage, runs east–west. That is to say, it lies in the vicinity of the ship's position along a parallel of latitude. It is for this reason that latitude is so readily found from an observation of a celestial body at meridian passage.

We have seen in Part I, Chapter V, that the latitude of an observer is equivalent to the altitude of the celestial pole. If a celestial body crosses the observer's upper or superior celestial meridian, the body is said to *culminate*, or to be at *superior transit* or *passage*, or to be *on the meridian above the pole.*

Celestial bodies which are circumpolar cross the meridian of an observer above the horizon on two occasions during each diurnal circuit. When such a body crosses the lower celestial meridian of an observer it is said to be at *lower* or *inferior transit*, or to be *on the meridian below the pole.* Let us deal separately with the problems of finding latitude from observations of celestial bodies at upper and lower transits.

LATITUDE FROM AN OBSERVATION OF A CELESTIAL BODY ON THE MERIDIAN ABOVE THE POLE

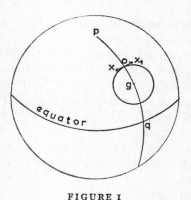

FIGURE I

Fig. 1 illustrates the Earth. p is the North Pole; o is an observer on meridian pq who observes a celestial object at upper meridian passage. g represents the geographical position of the observed object.

The small circle in Fig. 1 represents a circle of equal altitude centred at g and which passes through the observer's position. XX_1 is a small fragment of the circle of altitude through o. This, when projected on to the chart, will be the position line obtained from the observation. Because the observed body bears due south, the position line coincides with the parallel of latitude of the observer.

From Fig. 1:

arc qo = Latitude of observer

arc qg = Declination of observed object

arc go = M.Z.D. of the observed object, i.e. (90° — Meridian altitude of observed object).

Now qo = qg + og

therefore:

Latitude of observer = Declination of observed object + M.Z.D. of observed object

= Dec of object + (90° — M.Alt)

This relationship may be illustrated using the celestial sphere instead of the terrestrial sphere.

Fig. 2 illustrates the celestial sphere on the plane of the observer's celestial meridian. The small circle at the centre represents the Earth. p is the North Pole and P is the elevated celestial pole. o is an observer whose zenith is at Z. QQ_1 is in the plane of the equinoctial and N and S are the north and south points of the observer's horizon.

Fig. 3 illustrates the same conditions that pertain to Fig. 2, but it represents the celestial sphere drawn on the plane of the observer's horizon.

From Figs. 2 and 3:

arc NP = Altitude of celestial pole

= Latitude of observer

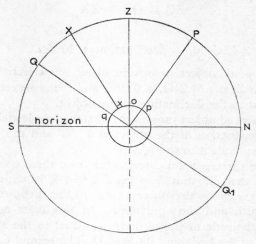

FIGURE 2

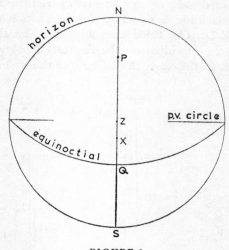

FIGURE 3

Because	NZ = 90°
and	PQ = 90°
therefore	ZQ = NP

But $$ZQ = QX + ZX$$

therefore:

$$\text{Latitude} = \text{Declination} + \text{M.Z.D.}$$

If the observed object crosses the observer's celestial meridian at the zenith Z, the M.Z.D. is 0° 00′, and the observer's latitude is equivalent to the declination of the object.

If the observed object crosses the observer's celestial meridian at Q, the declination of the object is 0° 00′ and the observer's latitude is equivalent to the M.Z.D.

There are three general cases, apart from these special cases. The first is that, illustrated in Figs. 2 and 3, in which the observed body crosses the observer's meridian between the observer's zenith and the equinoctial. In this case, as we have shown, the latitude of the observer is equal to the sum of the declination of the body and its M.Z.D. The second case applies when the observed object crosses the observer's meridian on the elevated poleward side of the observer's zenith. In this case the observer's latitude is equal to the declination of the body minus its M.Z.D. The third case applies when the observed object crosses the meridian on the depressed poleward side of the equinoctial. In this case, the latitude of the observer is equal

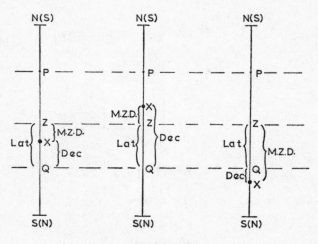

FIGURE 4

to the M.Z.D. of the object minus its declination. In all cases, the latitude of the observer is a combination of the M.Z.D. of an observed body and its declination. Fig. 4, which represents the celestial meridian of an observer on the plane of his horizon, illustrates the three cases.

There are many aids to memory designed to assist navigators who have little or no knowledge of the principles of the problems of combining declination and M.Z.D. to find latitude. The rules work if they are applied properly. But the principle is simple; and, if a rough drawing of the conditions is made, there is no need to resort to a mnemonic (or *donkey's bridge* as the Dutch call it).

LATITUDE FROM AN OBSERVATION OF A CELESTIAL BODY ON THE OBSERVER'S LOWER CELESTIAL MERIDIAN

A body which is visible at lower meridian passage is a circumpolar body. Conditions necessary for a celestial body to be circumpolar have been discussed in Part I, Chapter V.

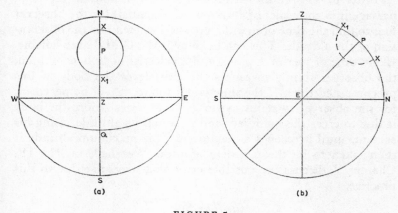

FIGURE 5

Fig. 5 illustrates the celestial sphere. Diagram (a) is drawn on the plane of the horizon of an observer whose zenith is at Z. Diagram (b) is drawn on the plane of the celestial meridian of the same observer.

P represents the celestial pole, and X and X_1, a celestial body at lower and upper meridian passage respectively.

When a circumpolar body is at lower meridian passage it attains its least altitude for the day. The radius of its diurnal circle is equivalent to the complement of its declination. This angle is generally called the *polar distance* (P.D.) of the object.

Because the altitude of the celestial pole is equivalent to the latitude of an observer, it follows (and this will readily be seen from Fig. 5) that:

Latitude of observer = Altitude of celestial body at lower meridian passage + P.D. of object

From Fig. 5:

$$NP = NX + PX$$

i.e.   Latitude = Altitude + P.D.

FINDING TIME OF MERIDIAN PASSAGE

In order to observe the altitude of a celestial body at meridian passage it is convenient (although not essential) for the observer first to find the time of meridian passage to facilitate the observation of the altitude. This can be done if the G.M.T. is available. If the time of meridian passage of a celestial body is computed, the observer simply measures the altitude of the body at this precomputed time. If the time is not found, it will be necessary for the observer to stand by for some minutes before the body is due to cross the meridian, and to watch its altitude, using his sextant, until it reaches a maximum. This maximum altitude is then taken to be the meridian altitude. We shall, in Part IV, Chapter V, discuss the possible error that may arise due to this practice.

a. The Sun

When the Sun is at upper meridian passage his L.H.A. is 00 hr and the L.A.T., therefore, is 12 hr 00 min. The observer's longitude measured west from the Greenwich meridian applied to the Sun's L.H.A. gives the Sun's G.H.A. for the time. This is tabulated in the *Nautical Almanac* against G.M.T. so that the G.M.T. of the Sun's meridian passage may readily be found.

An alternative method of finding the G.M.T. of the Sun's meridian passage is to use the tabulated time of meridian passage which is to be found in the daily page of the *Nautical Almanac*. These tabulated times are strictly G.M.T's of the Sun's meridian passages across the Greenwich celestial meridian, but they may be taken as being equivalent to L.M.T's of the Sun's meridian passages across local meridians. This follows because the Mean Solar Day is almost equal in length to that of the Apparent Solar Day. If the L.M.T. of the True Sun's meridian passage can be found, the L.M.T. of the Mean Sun's meridian passage can be found by applying the equation of time. On a day when the equation of time is say +14 min the L.M.T. of the True Sun's meridian passage would be 12 hr 14 min. In other words 12 hr 00 min L.A.T. corresponds to 12 hr 14 min L.M.T. on the day when the equation of time is +14 min. When the equation of time is negative L.M.T. is less than L.A.T., so that the L.M.T. of the Sun's upper meridian passage occurs before Local Apparent Noon when the equation of time is negative.

If the L.M.T. of an event is known, the longitude of the observer applied to it will give the G.M.T. of the event. When observing the Sun at upper meridian passage it is convenient, therefore, to work out the G.M.T. of the instant, and by the help of the chronometer to measure the altitude at this predetermined G.M.T.

Finding the G.M.T. of the meridian passage of the Moon, a planet or a star for the purpose of finding latitude by meridian altitude can hardly be regarded, in these enlightened days, to be a practical problem of navigation. There is generally no need to take the trouble to do so. This follows because an observation of any celestial body at any time will yield no more than a single position line: that obtained from an observation of a celestial body at meridian passage differs from other position lines only in respect of direction. However there may be an occasional circumstance (perhaps to demonstrate a principle of astronomical navigation to a cadet) when it might be useful to work out the G.M.T. of the meridian passage of a star, the Moon, or a navigational planet.

b. The Moon

The G.M.T's of the Moon's meridian passage (both upper and lower) are tabulated for the meridian of Greenwich on the daily pages of the *Nautical Almanac*. Now the interval between successive upper meridian passages of the Moon across any given observer's celestial meridian is always more than 24 hours of Mean Solar Time. For example, on February 10th 1968, the G.M.T. of the Moon's upper meridian passage is 21 hr 24 min. On the following day, it is 22 hr 20 min. The interval between these two times is 25 hr 04 min. Thus, the lunar day on this occasion is 64 min longer than a Mean Solar Day.

The G.M.T. of the Moon's transit on February 10th at the meridian of 90° W., may be found by adding 90/360th or a quarter of a lunar day to the G.M.T. of the Moon's meridian passage at Greenwich on February 10th. This will give an approximate, although practical, result, because of the invalid assumption that the Moon's diurnal motion on the celestial sphere is uniform.

To find the G.M.T. of the Moon's transit for an east longitude, a proportion of the lunar day (depending upon the longitude) is to be subtracted from the G.M.T. of the Moon's meridian passage at Greenwich for the day in question. The following examples will illustrate this.

EXAMPLE: Find the G.M.T. of the Moon's upper transit across the meridian of:

(a) 30° W. on February 10th 1968
(b) 120° E. on February 11th 1968

(a) From the *Nautical Almanac*:

	h	m
G.M.T. of ☽'s mer. pass. at Greenwich on 10th =	21	24
G.M.T. of ☽'s mer. pass. at Greenwich on 11th =	22	20
Length of lunar day =	25	04

Proportion for 30° = $\frac{30}{360}$ × 25 hr 04 min   = 02 05
G.M.T. ☾'s mer. pass. at long. 0° on 10th   = 21 24

∴ G.M.T. ☾'s mer. pass. at long. 30° W. on 10th   = 23 29

                                                                         h   m

(b) Proportion for 120° = $\frac{120}{360}$ × 25 hr 04 min   = 08 21
    G.M.T. ☾'s mer. pass. at long. 0° on 11th   = 22 20

∴ G.M.T. ☾'s mer. pass. at long. 120° E. on 11th = 13 59

An alternative method of solving the above problems involves first finding the L.M.T. of the Moon's meridian passage at ship, and then applying the longitude to this to give the required G.M.T.

The difference between the G.M.T. of the Moon's meridian passage at Greenwich and the L.M.T. of the Moon's meridian passage at ship is called the *longitude correction*.

The longitude correction is a proportion of the excess of the lunar day over 24 hr 00 min and is proportional to the longitude. The following solution to the examples given above will exemplify the alternative method of finding the G.M.T. of the Moon's transit.

                                                                      h   m

(a) G.M.T. ☾'s mer. pass. at long. 0° on 10th   = 21 24
    Longitude correction ($\frac{30}{360}$ × 64 min)   = +05

    L.M.T. ☾'s mer. pass. at long. 30° W. on 10th = 21 29
    Longitude   = 02 00

∴ G.M.T. ☾'s mer. pass. at long. 30° W. on 10th = 23 29

                                                                       h   m

(b) G.M.T. ☾'s mer. pass. at long. 0° on 11th   = 22 20
    Longitude correction ($\frac{120}{360}$ × 64 min)   = −21

    L.M.T. ☾'s mer. pass. at long. 120° E. 11th   = 21 59
    Longitude   = 08 00

∴ G.M.T. ☾'s mer. pass. at long. 120° E. 11th   = 13 59

c. A Planet

The G.M.T. of a navigational planet's upper meridian passage at Greenwich is given in the *Nautical Almanac* to the nearest minute of time for every third day. Because the day by the planet is seldom more or less by a few minutes than the length of a Mean Solar Day, the tabulated values are approximate L.M.T's of local meridian passage. In other words, the longitude correction in the case of a planet is trifling and, accordingly, in practice it is unworthy of consideration.

d. A Star

To facilitate finding the G.M.T. of a star's upper meridian passage, the G.M.T. of the transit of the First Point of Aries at Greenwich is tabulated in the *Nautical Almanac* at three-day intervals. Because the sidereal day is shorter than the Mean Solar Day by about four minutes of Mean Solar Time, the G.M.T. of the transit of the First Point of Aries across the Greenwich (or other) celestial meridian is later on succeeding days by this amount.

When a star is at upper meridian passage its R.A. is equivalent to the Local Sidereal Time (L.S.T.). This follows because L.S.T. at any instant is equivalent to the L.H.A. of the First Point of Aries at the instant; and the L.H.A. of the First Point of Aries is equivalent to the L.S.T.

The R.A. of a star may be found by subtracting its S.H.A. from 360°. By applying the observer's longitude to a star's R.A. the result is the G.H.A. of the First Point of Aries at the time of the star's meridian passage. Fig. 6 illustrates this.

In Fig. 6, P represents the north celestial pole. PG, PO and P♈ represent respectively the celestial meridians of Greenwich, an observer, and the hour circle of the First Point of Aries. X is a star at meridian passage.

From Fig. 6:

$$\text{arc } G♈ = \text{arc } O♈ + \text{arc } GO$$

i.e.   G.H.A. ♈ = L.H.A. ♈ + W. longitude of observer

Because G.H.A. ♈ is tabulated in the *Nautical Almanac*

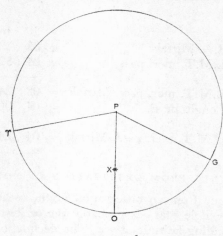

FIGURE 6

against G.M.T., if the G.H.A. corresponding to a star's meridian passage can be found, the G.M.T. of the star's meridian passage can be found by interpolation using the interpolation tables provided in the *Nautical Almanac*.

In practice it is not necessary to go to this trouble. By treating the G.M.T. of the Greenwich transit of the First Point of Aries as being equivalent to the L.M.T. of the local transit of the First Point of Aries, the problem is facilitated by adding the R.A. of the star to the G.M.T. of the transit of the First Point of Aries to find the L.M.T. of local transit. The longitude is applied to this to obtain the required G.M.T. The following example illustrates this:

EXAMPLE: Find the G.M.T. of the upper transit of the star Mirfak over the meridian of 60° E. on July 10th 1968.

From the *Nautical Almanac*:

		° ′
S.H.A. Mirfak	=	309 29·5
	=	360 00
R.A. Mirfak	=	50 30·5

In time

		h	m
R.A. Mirfak	=	03	22
L.M.T. mer. pass. ♈	=	04	52
L.M.T. mer. pass. Mirfak	=	08	24
Longitude E.	=	04	00
G.M.T. mer. pass. Mirfak	=	04	24

LATITUDE FROM OBSERVATION OF POLARIS

Finding latitude from an observation of the 'seaman's star'—Polaris, or the Pole Star—is probably the earliest astronomical method for finding latitude at sea. The declination of the relatively bright star α Ursae Minoris (magnitude 2·2) is a little more than 89° (89° 06′ in 1965). It lies, therefore, within a degree of the north celestial pole. It follows that because the altitude of the celestial pole is equal to the latitude of the observer, the altitude of the Pole Star is never different from the latitude of the observer by a degree or less. When Polaris is on the observer's upper celestial meridian its altitude is about one degree greater than the latitude of the observer. When it is at lower meridian passage its altitude is about one degree less than the latitude of the observer. When its L.H.A. is about 06 hours or 18 hours its altitude is roughly equal to the latitude of the observer.

The correction to apply to the altitude of Polaris, in order to find the latitude of the observer, is provided in the *Nautical Almanac*. The correction is a function of the L.H.A. of the star and the latitude of the observer. The Pole Star Tables in the *Nautical Almanac* give three corrections against arguments L.H.A. ♈, latitude of observer, and month of the year. The three corrections are denoted by a_0, a_1 and a_2. They are to be added together and their sum diminished by 1°. The latitude of the observer is then found by applying the resultant correction to the True Altitude of the star. Thus:

$$\text{Latitude} = \text{True Altitude} - 1° + a_0 + a_1 + a_2$$

In addition to the tables for finding the latitude from an

altitude observation of Polaris, an Azimuth Table is also given in the *Nautical Almanac*.

The Pole Star Tables provide a convenient method for finding latitude when the observer is between about 10° and 68° N. South of about 10° N. Polaris is too near the horizon for it to be suitable for navigational purposes. The tables extend only to 68° N., which latitude approximates to the northern limit of surface navigation.

The formula used in calculating Pole Star Tables is:

$$\text{Correction to alt} = -p \cos h + p/2 . \sin p \sin^2 h \tan \phi$$

where p = P.D. of Polaris

   h = L.H.A. of Polaris

   = L.H.A. ♈ + S.H.A. Polaris

   ϕ = observer's latitude (which is very nearly equal to the altitude a of Polaris)

The derivation of the Pole Star formula is described with reference to Fig. 7.

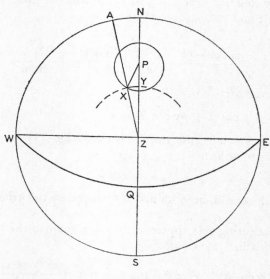

FIGURE 7

Fig. 7 represents the celestial sphere on the plane of the horizon of an observer whose zenith is at Z and whose latitude is ϕ which is equivalent to arc NP. The small circle represents the diurnal path of Polaris (greatly exaggerated for diagram purposes) and X denotes Polaris at the instant its L.H.A. is h.

Let the correction to be applied to the altitude AX (denoted by a) to find the latitude NP (denoted by ϕ) be y. This is represented by arc PY in Fig. 7. Then:

$$\text{Latitude} = \text{Altitude} - \text{correction}$$

i.e. $$\text{Latitude} = a - y$$

Applying the spherical cosine formula to the astronomical triangle PZX we have:

$$\cos h = \frac{\cos ZX - \cos PZ \cos PX}{\sin PZ \sin PX}$$

i.e. $$\cos h = \frac{\sin a - \sin (a - y) \cos p}{\cos (a - y) \sin p}$$

Since y and p are small quantities, we may assume the equivalence of y and $\sin y$ and p and $\sin p$, and we may call $\cos y$, $(1 - y^2/2)$; and $\cos p$, $(1 - p^2/2)$.

Thus:

$$\cos h = \frac{\sin a - (1 - p^2/2)\{\sin a(1 - y^2/2) - y \cos a\}}{p \cos a(1 - y^2/2) + py \sin a}$$

This reduces to:

$$y = p \cos h + py \cos h \tan a - \tan a(p^2 + y^2)/2 \qquad (1)$$

Neglecting second-order terms in equation (1) we get as a first approximation:

$$y = p \cos h$$

This result would be obtained by assuming the triangle PXY to be plane.

By substituting this approximate value for y in the right-hand side of equation (1), we get:

$$y = p \cos h + p^2 \cos^2 h \tan a - \tan a\left(\frac{p^2 + p^2 \cos^2 h}{2}\right)$$

i.e.

$$y = p \cos h - \tan a \sin^2 h . p^2/2$$

But $a = \phi$ and $p = \sin p$, therefore:

$$y = p \cos h - \tan \phi \sin^2 h \sin p . p/2$$

Because the correction is negative the correction to the altitude to find the latitude is:

$$- p \cos h + p/2 \sin p \sin^2 h \tan \phi$$

Values of a_0 in the Pole Star Tables are computed using both terms of the formula.

The first term in the formula, which depends upon L.H.A. (and therefore on S.H.A. of Polaris) and the polar distance of the star, is computed using mean values for S.H.A. and polar distance. (These for 1968 are 329° 40′ and 89° 07·3′ N. respectively.)

The second term in the formula is computed using the same mean values for S.H.A. and P.D. of Polaris, and for a mean value of 50° for the latitude. The combination of these terms are then adjusted by the addition of a constant (58·8′) so that the values of a_0 are always positive.

Values of a_1 in the Pole Star Tables depend upon L.H.A. and the observer's latitude. They represent the excess of the value of the second term over its mean value for latitude 50°, increased by a constant (0·6′) to make the correction always positive.

Values of a_2 take into account the variations of S.H.A. and P.D. of Polaris during the year. These values depend upon the L.H.A. ♈ and the date. They are increased by a constant (0·6′) so that they are always positive.

It will be noticed that the sum of the three constants used to adjust the three corrections to get a_0, a_1 and a_2, is exactly 1°. Thus:

$$\text{Latitude} = \text{Altitude} + (a_0 + a_1 + a_2) - 1°$$

It is customary first to subtract 1° from the True Altitude and then to add the three corrections a_0, a_1 and a_2.

The Pole Star Azimuth Table gives azimuths correct to 0·1°. This relatively coarse degree of accuracy (which is still finer than that required for practical navigation) suggests that refinements like those used for computing the correction to find latitude are not used in computing tabulated azimuths.

Referring to Fig. 7:

In the astronomical triangle PZX:

$$\text{Azimuth of Polaris at X} = PZX$$
$$= \text{arc NA}$$

By the parallel sailing formula:

$$\text{arc NA} = \text{arc XY sec AX}$$

therefore: $\text{Azimuth} = XY \sec a$

but $XY = p \sin h$ (approx.)

therefore: $\text{Azimuth} = p \sin h \sec a$

An alternative, but more complex, method of dealing with the azimuth of Polaris is to apply the spherical four-parts formula to the PZX triangle as follows:

$$\sin \phi \cos h = \cos \phi \cot p - \sin h \cot Z$$

from which:

$$\tan Z = \frac{\sin h}{\cos \phi \cot p - \sin \phi \cos h}$$

that is,

$$\tan Z = \frac{\sin h \tan p}{\cos \phi - \sin \phi \cos h \tan p}$$

Because Z and p are small quantities we may assume the equivalence of tan Z and Z; and sin h and h. Therefore

$$Z = \frac{p \sin h}{\cos \phi - p \sin \phi \cos h}$$

that is:

$$Z = p \sin h(\cos \phi - p \sin \phi \cos h)^{-1}$$

Expanding the right-hand side of this formula using the

binomial theorem, and neglecting terms higher than the second power of p, we get:

$$Z \text{ (rad)} = p \sin h \sec \phi(1 + p \tan \phi \cos h)$$

or,

$$Z \text{ (min)} = 3438 \, p \sin h \sec \phi(1 + p \tan \phi \cos h)$$

LATITUDE BY EX-MERIDIAN ALTITUDE OBSERVATION

The term 'ex-meridian' in this context means 'near the meridian'.

When a celestial object is near meridian passage the observer may find from an altitude observation the latitude of his ship by using a method which has become known as *Latitude by Ex-meridian*. This method of finding latitude, like that of the meridian altitude method, no longer holds its former importance. It is mainly of historical interest, but candidates for professional examinations, requiring knowledge of the method, will find the following account of value.

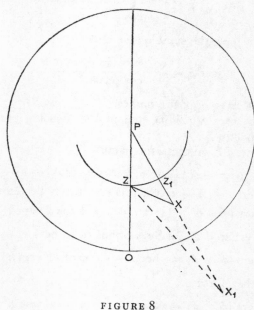

FIGURE 8

The essential problem in the ex-meridian method for finding latitude is the comparison of the altitude of a celestial body at a place where the body is culminating (the latitude of the place being the same as the observer's latitude), with its altitude for the same instant of time at the observer's actual, but unknown, position. Fig. 8 serves to illustrate the ex-meridian problem.

Fig. 8 illustrates the celestial sphere drawn on the plane of the horizon of an observer whose zenith is at Z. PO represents the observer's meridian. Z_1 is the projection of the zenith of a place the latitude of which is the same as that of the observer's, and over whose meridian the body X is passing.

If the arc Z_1X can be found, the latitude of the place whose zenith is at Z_1, and hence the observer's latitude, may also be found.

If ϕ, d, z and h denote the observer's latitude, the body's declination, the body's zenith distance, and the time from meridian passage of the body, respectively, we have from the astronomical triangle PZX:

$$\cos h = \frac{\cos z - \sin \phi \sin d}{\cos \phi \cos d}$$

when ϕ and d have the same name, and

$$\cos h = \frac{\cos z + \sin \phi \sin d}{\cos \phi \cos d}$$

when ϕ and d have opposite names.

These two cases are illustrated in Fig. 8 using body X and body X_1 respectively.

When ϕ and d have the same name:

$$\cos z - \sin \phi \sin d = \cos h \cos \phi \cos d$$

Because vers $\theta = 1 - \cos \theta$, this result may be reduced to:

$$\text{vers}\,(\phi \sim d) = \text{vers}\,z - \cos \phi \cos d \,\text{vers}\,h$$

Similarly, when ϕ and d have opposite names:

$$\text{vers}\,(\phi + d) = \text{vers}\,z - \cos \phi \cos d \,\text{vers}\,h$$

In general, therefore:

$$\text{vers}\,(\phi \pm d) = \text{vers}\,z - \cos \phi \cos d \,\text{vers}\,h$$

Also, because haversine $\theta = \frac{1}{2}$ versine θ, therefore:

$$\text{hav} \ (\phi \pm d) = \text{hav} \ z - \cos \phi \cos d \ \text{hav} \ h$$

Now $(\phi \pm d)$ is the M.Z.D. of the body X or X_1 at the place whose zenith is at Z_1. The latitude of this place, and, therefore, the observer's latitude, is thus given by:

$$\phi = \text{M.Z.D.} \pm d$$

This method of finding the latitude of the observer requires the use of an estimated latitude which should approximate to the observer's actual, but unknown, latitude. If the latitude found differs materially from that used, it is necessary to repeat the computation, this time using the calculated latitude instead of the estimated latitude used in the first computation. Moreover, it is necessary for the observer to use an estimated longitude— knowledge of this being necessary to find h which figures in the computation.

It was early realized that when using stars for finding latitude by the ex-meridian method, those with big declinations give the best results (see Part II, Chapters V and VI). This follows because of their relatively slow rates of change of altitude. The Pole Star is admirably suited for this purpose, and the Pole Star tables we have described are, in a sense, ex-meridian tables for this body.

Of theoretical interest is a method of finding latitude from an ex-meridian altitude observation using right-angled spherical trigonometry. This is described with reference to Fig. 9.

Fig. 9 illustrates the celestial sphere on the plane of the horizon of an observer whose zenith is at Z. P is the elevated celestial pole and X is a star whose hour angle is h. Point Y, on the observer's celestial meridian, lies on the great circle through X which is perpendicular to the observer's celestial meridian.

In the triangle PYX:

$$\tan PY = \tan PX \cos P \tag{1}$$

$$\cos PX = \cos YX \cos PY \tag{2}$$

In the triangle XYZ:

$$\cos ZX = \cos ZY \cos YX$$

from which, using (2), we can obtain:

$$\cos ZY = \cos ZX \cos PY \sec PX \qquad (3)$$

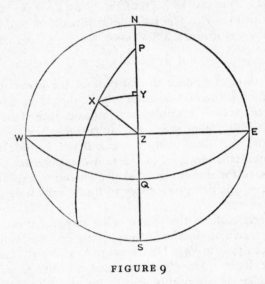

FIGURE 9

From equations (1) and (3) the arcs PY and ZY may be found. These, when combined, give arc PZ which is the complement of the observer's latitude.

This method is independent of the latitude and may be used to good effect even when the body's hour angle is large provided that it is known accurately. This will depend on knowledge of the longitude.

A method alternative to the direct methods, like those described above, is known as the *Reduction to the Meridian* method. This involves the computation of a correction to apply to the ex-meridian zenith distance to find the M.Z.D. The reduction method makes use of an estimated latitude which, if materially different from the calculated latitude, requires recalculation of the problem using the calculated latitude found in the first calculation.

The reduction to the meridian method is described with reference to Fig. 10.

Fig. 10 represents the celestial sphere drawn on the plane of the horizon of an observer whose zenith is at Z. X is the projection of a celestial body whose hour angle at the time of the observation is h, and whose declination and zenith distance are d and z respectively. Arc ZA is equal to arc XZ.

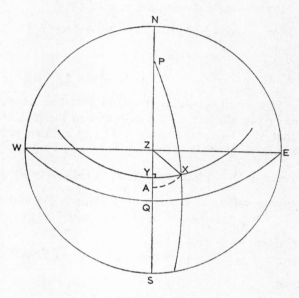

FIGURE 10

The arc YA is the reduction to the meridian. This is clearly the difference between the zenith distances of X at the times of observation and meridian passage respectively. If the reduction is denoted by r, then:

$$r = \text{Ex-M.Z.D.} - \text{M.Z.D.}$$

The ex-meridian problem has provided a favourite field of inquiry for investigators into the problems of nautical astronomy. The fruits of these investigations are prolific. There are dozens of methods of finding latitude by the ex-meridian principle and there are scores of ex-meridian tables designed to facilitate finding latitude at sea.

The reduction method is usually attributed to the great French astronomer Delambre, who published his method in 1814. Delambre applied the spherical cosine formula to the astronomical triangle PZX as follows:

In the triangle PZX:

$$\cos z = \cos h \cos \phi \cos d + \sin \phi \sin d$$

i.e. $$\cos z = \cos(\phi \pm d) - 2 \cos \phi \cos d \sin^2 h/2$$

Now, $$z = (ZY + r)$$

and, $$\cos(ZY + r) = \cos ZY \cos r - \sin ZY \sin r$$

Since r is a small quantity, provided that the astronomical conditions favour the use of the ex-meridian method, we may assume that $\sin r = r$; and $\cos r = 1 - r^2/2$. Also since arc ZY = M.Z.D. = $(\phi \pm d)$, it follows that:

$$\cos z = (1 - r^2/2) \cos (\phi \pm d) - r \sin (\phi \pm d) \qquad (2)$$

By equating values of $\cos z$ from equations (1) and (2) we have:

$$(1 - r^2/2) \cos (\phi \pm d) - r \sin (\phi \pm d)$$
$$= \cos (\phi \pm d) - 2 \cos \phi \cos d \sin^2 h/2$$

from which:

$$r^2/2 \cos (\phi \pm d) + r \sin (\phi \pm d)$$
$$= 2 \cos \phi \cos d \sin^2 h/2 \qquad (3)$$

The first term in (3) is small when the body is near meridian passage. It may, therefore, be neglected; and, for practical purposes:

$$r = 2\left(\frac{\cos \phi \cos d}{\sin (\phi \pm d)}\right) \sin^2 h/2$$

Now, $$\sin^2 \theta/2 = \text{hav } \theta$$

therefore:

$$r = 2\left(\frac{\cos \phi \cos d}{\sin (\phi \pm d)}\right) \text{hav } h$$

If r is expressed in minutes of arc, then:

$$r' = 3438 \times \left(\frac{2 \cos \phi \cos d}{\sin (\phi \pm d)}\right) \text{hav } h$$

If r is expressed in seconds of arc:

$$r'' = 60 \times 6876 \left(\frac{\cos \phi \cos d}{\sin (\phi \pm d)}\right) \text{hav } h$$

The ex-meridian tables found in Norie's and Burton's collections give values of the change in altitude in seconds of arc during one minute of time from or to that of meridian passage. This is tabulated as A (or F) so that:

$$A'' = 60 \times 6876 \times \left(\frac{\cos \phi \cos d}{\sin (\phi \pm d)}\right) \text{hav 1 min}$$

$$= 1 \cdot 9635 \times \frac{\cos \phi \cos d}{\sin (\phi \pm d)}$$

The arguments used in Table I of the ex-meridian tables are latitude and declination.

Since the average rate of change of altitude during the period of one minute from or to the instant of meridian passage is A'', the rate at the time of meridian passage being zero, means that the rate at a minute before or after the time of meridian passage must be $2A$ seconds of arc per minute of time.

The change in altitude during the minute before or after the time of meridian passage is considered to be an acceleration; and, from the relationship between distance s travelled at uniform acceleration a in time t, viz, $s = \frac{1}{2}at^2$, we get:

Change in altitude in h min $= Ah^2$

Table II of the ex-meridian tables in Norie's and Burton's collections performs the multiplication of A and h^2. The arguments used in the table are, accordingly, A and time from or to that of meridian passage.

The result obtained from Table II is an approximation of the reduction to the meridian. This, subtracted from the observed zenith distance gives the M.Z.D. from which a latitude may be found by applying the body's declination. The latitude obtained is the ship's latitude only if the longitude used in finding it is the ship's longitude. Because, in general, the longitude of the ship is not known at the time, the latitude obtained is not the ship's latitude. So that, strictly speaking, the result of an ex-meridian observation is not the ship's latitude but a position line which passes through the calculated latitude and the longitude used in the calculation.

Table III of the ex-meridian tables of Norie's and Burton's should be used when h is sufficiently large to make the result, using Tables I and II only, insufficiently accurate. Table III, therefore, may be regarded as providing the means of extending the use of Tables I and II.

Referring back to Fig. 9.

$$\cos z = \sin \phi \sin d + \cos \phi \cos d \cos h$$

$$= \cos (\phi \pm d) - 2 \cos \phi \cos d \sin^2 h/2 \qquad (1)$$

Now
$$z = \{(\phi \pm d) + r\}$$
therefore:

$$\cos z = \cos (\phi \pm d)\{1 - 2 \sin^2 r/2\} - \sin (\phi \pm d) \sin r \qquad (2)$$

By equating (1) and (2) we get:

$$\cos (\phi \pm d) - 2 \cos \phi \cos d \sin^2 h/2$$
$$= \cos (\phi \pm d)\{1 - 2 \sin^2 r/2\} - \sin (\phi \pm d) \sin r$$

from which:

$$\sin r = \frac{2 \cos \phi \cos d \text{ hav } h}{\sin (\phi \pm d)} - 2 \sin^2 r/2 \cot (\phi \pm d)$$

If r is small and it is expressed in minutes of arc, we have:

$$\frac{r}{3438} = \frac{2 \cos \phi \cos d \text{ hav } h}{\sin (\phi \pm d)} - \cos (\phi \pm d)2\left(\frac{r}{3438}\right)^2$$

i.e.

$$r = \frac{6876 \cos \phi \cos d \operatorname{hav} h}{\sin (\phi \pm d)} - \cot (\phi \pm d) \frac{2r^2}{3438}$$

The second term in this expression forms the basis of the values tabulated in Table III of the ex-meridian tables.

$$\text{Additional correction} = - \cot (\phi \pm d)\frac{2r^2}{3438} \text{ min of arc}$$

$$= - \cot (\phi \pm d)\frac{r^2}{28 \cdot 65} \text{ sec of arc}$$

Now $(\phi \pm d) = z$, and z is the complement of the altitude of the observed body. Therefore, additional correction is:

$$- \tan a \frac{r^2}{28 \cdot 65} \text{ sec of arc}$$

In the above discussion, we have used the relatively loose term 'near the meridian'. It now remains to define this term with some measure of precision.

The term 'near the meridian' implies an hour angle limited according not only to latitude and declination, but also to the accuracy with which G.M.T. is known and to the required degree of accuracy of the calculated latitude.

It is a common practice to provide a table (ex-meridian Table IV in Norie's collection) giving limits of time from or to meridian passage computed to give the number of minutes in the hour angle when an error of a given amount in the hour angle produces an error of a given amount in the reduction. The usual table gives limits of hour angle when an error of half a minute in time in the hour angle produces an error of one minute of arc in the zenith distance and hence in the calculated latitude.

The practical rule using this relationship in respect of the Sun is:

'The number of minutes to or from apparent noon should not exceed the number of degrees in the Sun's zenith distance.'

N.B. A final note to the old-fashioned navigators who still pin their faith in the ex-meridian method: Latitude found from an ex-meridian observation of the Sun is the latitude of a point on a position line the direction of which is at right angles to the bearing of the Sun at the time of the observation. IT IS NOT, REPEAT NOT, THE LATITUDE OF THE SHIP AT NOON.

CHAPTER V

Rates of Change

In this chapter we shall be concerned primarily with the accelerations of a celestial body during its diurnal circuit relative to an observer's celestial meridian and the body's vertical circle respectively.

The arc of the horizon contained between the vertical circle through the elevated pole and that through any celestial body is a measure of the body's azimuth. The arc of a vertical circle contained between a celestial body and the horizon vertically beneath the body is a measure of the altitude of the body. Both the azimuth and the altitude of a celestial body, except in special circumstances, change at rates that are not uniform. It is with the rates of change of azimuth and altitude that we shall be concerned.

The average rate of change of azimuth of a celestial body is a measure of the ratio between the change in the body's azimuth in any given interval of time t, and the interval of time itself. Thus:

Average rate of change of azimuth

$$= \frac{\text{change of azimuth in interval } t}{t}$$

In all cases, a body which changes its azimuth does so in consequence of a change in its hour angle, so that the expression may be written thus:

Average rate of change of azimuth

$$= \frac{\text{change of azimuth in } t}{\text{change of hour angle in } t} \times \frac{\text{change in H.A. in } t}{t}$$

The average rate of change of altitude of a celestial body is a measure of the ratio between the change in the altitude of the body in any given interval of time t, and the interval of time itself. Thus:

Average rate of change of altitude

$$= \frac{\text{change in altitude in interval } t}{t}$$

In all cases a celestial body which changes its altitude does so in consequence of a change in its hour angle, so that the expression may be written thus:

Average rate of change of altitude

$$= \frac{\text{change in altitude in } t}{\text{change in H.A. in } t} \times \frac{\text{change in H.A. in } t}{t}$$

If all celestial bodies changed their hour angles at the same rate as that of the Mean Sun, the change in hour angle in any given interval of time measured in Mean Solar units would be equal to the interval. This however is not the case.

To a stationary observer the rate of change of the Mean Sun's hour angle is 15° per Mean Solar hour, or 15′ per minute of time, or 15″ per second. This rate is uniform at all times.

Because the True Sun moves in the ecliptic his changing declination influences his rate of change of hour angle. This rate is irregular. The average rate of change of the True Sun's hour angle during a year is equal to the rate of change of the Mean Sun's hour angle. Because the variation from the average rate of change of the True Sun's hour angle is very small, his rate of change of hour angle is generally assumed to be the same as that of the Mean Sun. No material error is introduced by making this assumption.

The R.A. of the Mean Sun increases uniformly at the rate of 24 hours of Mean Solar time per year. This is equivalent to 2·46′ per hour. It follows that the rate of change of the hour angle of the First Point of Aries (or that of any fixed celestial point or star) is (900 + 2·46)′ per hour, or very nearly 902·5′ per hour.

The rate of change of hour angle of the Moon or any planet is very irregular. The value for any given time may readily be found from the *Nautical Almanac*.

If the rate of increase of the R.A. of a celestial body is r' per hour, the rate of change of its hour angle is $(902 \cdot 5 - r)'$ per hour.

Any motion of an observer on the Earth's surface, unless it be along a meridian, will influence the rate of change of a celestial body's hour angle. If, for example, an observer is moving towards the west at the same rate as the Earth is rotating towards the east, fixed celestial bodies would appear stationary in the heavens. It follows that any movement of an observer towards the west reduces the rate of change of hour angle of any celestial body compared with its rate when the observer is stationary. Conversely, any movement of an observer towards the east increases the rate of change of a body's hour angle compared with its rate for a stationary observer.

If the rate of change of longitude of a moving observer towards the west is x' per hour, the rate of change of hour angle of any celestial body is given by the expression:

Rate of change of hour angle $= (902 \cdot 5 - r - x)'$ per hour

Having derived an expression for the rate of change of hour angle of any celestial body, we are now in a position to investigate the rates of change of azimuth and altitude of celestial bodies.

RATE OF CHANGE OF AZIMUTH

The rate of change of a body's azimuth will be investigated with reference to Fig. 1.

Fig. 1 illustrates the celestial sphere drawn on the plane of the horizon of a stationary observer whose zenith is projected at Z. P is the projection of the celestial pole and N, E, S and W, that of the principal points on the observer's horizon.

Let us suppose that a celestial body of constant declination d, in travelling along its diurnal path from X to Y, changes its hour angle by one minute of time, denoted by the angle APB or the arc AB. In so doing the body changes its azimuth by the angle CZD or arc CD, and its altitude by angle (CX − DY), which is equivalent to arc VY—the point V lying on the same parallel of altitude as that through X. Let the altitude of the body when at X be denoted by α.

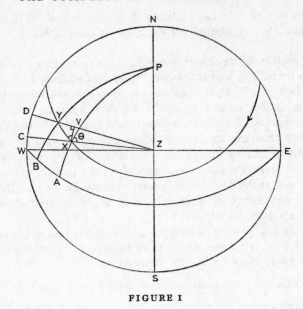

FIGURE I

Because arc XY is small, we may assume that the triangle XYV is plane without introducing material error.

Now	$YVX = 90°$
and	$YXP = 90°$
therefore	$YXV = 90° - VXP$
Also	$PXZ = 90° - VXP$
therefore	$YXV = PXZ$

Let this angle be denoted by θ.

The angle at a celestial body X in an astronomical triangle PZX(θ) is called the parallactic angle or angle of position.

$$
\left.
\begin{aligned}
\text{Rate of change of azimuth} &= CD \\
&= XV \sec \alpha \\
&= XY \cos \theta \sec \alpha \\
&= AB \cos d \cos \theta \sec \alpha
\end{aligned}
\right\}
\begin{aligned}
&\text{per} \\
&\text{minute} \\
&\text{of} \\
&\text{time}
\end{aligned}
$$

Now AB is one minute of time and this is equivalent to 15 minutes of arc. Therefore:

Rate of change of azimuth = $(15 \cos d \cos \theta \sec \alpha)'$ per minute

or = $(15 \cos d \cos \theta \sec \alpha)''$ per second

The above result may be obtained by using the differential calculus as follows:

Let δZ denote the change in azimuth (arc CD) consequent upon a change in hour angle of δh (arc AB). Then:

$$\text{Rate of change of azimuth} = \frac{\delta Z}{\delta h}$$

Now $\qquad XV = \delta Z \cos \alpha$

also $\qquad XV = XY \cos \theta$

$\qquad\qquad\qquad = \delta h \cos d \cos \theta$

therefore

$$\delta Z \cos \alpha = \delta h \cos d \cos \theta$$

and $\qquad \dfrac{\delta Z}{\delta h} = \sec \alpha \cos d \cos \theta$

In the limit as $\delta Z \to 0$, $\delta h \to 0$. Therefore:

$$\text{Limit } \frac{\delta Z}{\delta h} = \frac{dZ}{dh} = \sec \alpha \cos d \cos \theta$$

Expressing this rate of change of azimuth in minutes of arc per minute of time we have:

$$\text{Rate} = (15 \sec \alpha \cos d \cos \theta)' \text{ per minute}$$

Examination of this expression reveals that for a given declination the rate of change of azimuth of a heavenly body varies as the secant of the altitude and the cosine of the parallactic angle. For any given altitude the rate of change of azimuth of a heavenly body is greatest when cosine θ is a maximum, that is when $\theta = 0°$ or $180°$. This is so when the body is at upper or lower meridian passage respectively. It follows that a celestial body changes its azimuth most rapidly when it culminates and has its greatest altitude during its diurnal circuit.

12

When a celestial body is at upper meridian passage its zenith distance is equivalent to the sum or difference of the observer's latitude (ϕ) and the body's declination (d). That is to say:

$$\text{M.Z.D.} = (\phi \pm d)$$

The parallactic angle of a body at upper meridian passage is $0°$. It follows that the greatest rate of change of azimuth of a celestial body is given by:

$$\text{Rate} = [15 \cos d \operatorname{cosec} (\phi \pm d)]' \text{ per minute}$$

When a celestial body is at lower meridian passage its zenith distance is a maximum and is equivalent to $(90 - \phi) + (90 - d)$, that is $180 - (\phi + d)$. The parallactic angle of a body at lower meridian passage is $180°$. It follows that the rate of change of azimuth of a celestial body at lower transit is:

$$\text{Rate} = [15 \cos d \operatorname{cosec} (\phi + d)]' \text{ per minute}$$

From the general formula for rate of change of azimuth, viz.:

$$\text{Rate} = (15 \sec \alpha \cos d \cos \theta)' \text{ per minute}$$

it may readily be seen that when the parallactic angle is $90°$ the rate of change of azimuth is zero.

When the parallactic angle is $90°$;

$$\cos z = \frac{\sin \phi}{\sin d}$$

that is

$$\sin \alpha = \frac{\sin \phi}{\sin d}$$

Since sine α cannot be greater than unity, the ratio between $\sin \phi$ and $\sin d$ cannot be greater than 1. It follows that $\sin \phi$ must be less than $\sin d$ so that the conditions necessary for the parallactic angle to be $90°$ are that the observer's latitude must be less than the body's declination and the names of the latitude and declination must be the same.

When the parallactic angle is $90°$ the body is said to be at its *limiting azimuth*. When such a body is east or west of the ob-

server's celestial meridian its azimuth increases before, and decreases after, it reaches its limiting azimuth.

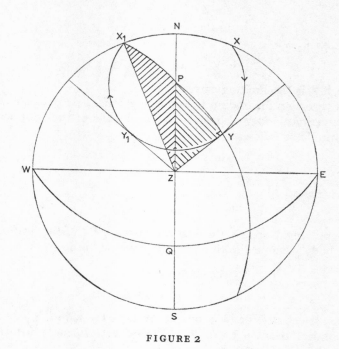

FIGURE 2

From Fig. 2, which illustrates the celestial sphere on the
plane of the horizon of an observer whose zenith is projected
at Z, it may readily be seen that the azimuth of a celestial body
which rises at X increases until the body reaches Y after which
the azimuth decreases to zero when the body is at meridian
passage. After the time of meridian passage the azimuth increases to a maximum which is reached when the body is at
Y_1 after which it decreases.

The points Y and Y_1 are the points in the diurnal circuit of the
body at which it is at limiting azimuth. When at Y or Y_1, the
body moves momentarily directly along the vertical circle it
occupies. That is to say, it changes its altitude but not its azimuth.

By applying the spherical sine formula to the triangle PZY

in Fig. 2, an expression for finding the limiting azimuth may readily be found:

$$\frac{\sin Z}{\sin PY} = \frac{\sin 90°}{\sin PZ}$$

From which:

$$\sin Z = \cos d \sec \phi$$

in which Z is the limiting azimuth.

An expression for finding the hour angle of a celestial body at limiting azimuth may also be found by means of the spherical sine formula as follows:

$$\frac{\sin P}{\sin ZY} = \frac{\sin 90°}{\sin PZ}$$

From which:

$$\sin P = \cos \alpha \sec \phi$$

By applying Napier's rules to the triangle PZY, the altitude of the body when at limiting azimuth may be found from:

$$\cos \alpha = \cos \phi \sin h$$

or

$$\cos \alpha = \cot d \cot Z$$

When a celestial body is on the horizon the side ZX of the astronomical triangle is 90°. In the general formula for the rate of change of azimuth viz.

$$\text{Rate} = (15 \sec \alpha \cos d \cos \theta)' \text{ per minute}$$

when $\alpha = 0°$, $\sec \alpha = 1$, so that for any celestial object on the horizon, its rate of change of azimuth is

$$(15 \cos d \cos \theta)' \text{ per minute}$$

Applying Napier's rules (see Appendix I, page 315) to the triangle PZX illustrated in Fig. 2, it may readily be shown that:

$$\cos d \cos X = \sin \phi$$

But X is the parallactic angle θ, therefore the rate of change of azimuth of a celestial body on the horizon is

$$(15 \sin \phi)' \text{ per minute}$$

Notice that the rate of change of azimuth of a body on the horizon is independent of the body's declination. When ϕ is 90° the rate of change of azimuth of a celestial body is 15′ per minute, and this rate is uniform and constant.

To an observer at either pole, because Z and P coincide, the parallactic angle at every celestial body is 0°. PZ and PX coincide, so that the rate of change of azimuth of every fixed celestial body to an observer in latitude 90° is 15′ per minute or 15° per hour.

Because sin 0° = 0, in latitude 0° the rate of change of azimuth of a celestial body on the horizon is zero. It follows that at the equator all celestial bodies rise out of and set into the horizon vertically.

RATE OF CHANGE OF ALTITUDE

Referring to Fig. 1, and again assuming conditions for a stationary observer and a celestial body of constant declination; the body, in the interval of time one minute (indicated by angle APB), changes its altitude by an amount equivalent to arc VY.

Because arc XY is small the triangle XYV may be assumed to be a plane. During the one-minute interval between the instants when the body is at X and Y respectively:

$$\text{Rate of change of altitude} = \frac{\text{change in altitude}}{t}$$

$$= \frac{VY}{AB}$$

$$= \frac{VY}{XY \sec d}$$

$$= \sin \theta \cos d \qquad (1)$$

It follows that when a body is at its limiting azimuth ($\theta = 90°$), its rate of change of altitude is maximum and is proportional to the cosine of its declination. When a celestial body is on an observer's celestial meridian $\theta = 0°$, so that the body's rate of change of altitude is zero. For a short period before and after the time of meridian passage sine θ decreases and increases respectively. Because for small angles sine $\theta \propto \theta^c$, it follows that for a short period before the time of meridian passage the change of altitude is a motion of uniform deceleration, and for a short period of time after the time of meridian passage, it is one of uniform acceleration.

By applying the spherical sine formula to the astronomical triangle PZX, we have:

$$\frac{\sin PX}{\sin Z} = \frac{\sin PZ}{\sin X}$$

from which

$$\sin \theta \cos d = \cos \phi \sin Z$$

By substituting $\cos \phi$ for $\sin \theta \cos d$ in (1), we have:

$$\text{Rate of change of altitude} = \cos \phi \sin Z$$

Expressing this rate in terms of minutes of arc per minute of time, we have:

Rate of change of altitude
$$= (15 \cos \phi \sin Z)' \text{ per minute} \qquad (2)$$

It follows from this formula that for any given latitude the rate of change of altitude of a heavenly body is greatest when sine Z is greatest, that is when Z is 90° or 270°. In other words, a body's rate of change of altitude is greatest for any given latitude when the body is on the prime vertical circle. The conditions necessary for a celestial body to cross the prime vertical circle are that its declination must be of the same name as but its magnitude smaller than that of the observer's latitude.

For a celestial body which does not cross the prime vertical circle and whose declination is different in name from that of the observer's latitude, the greatest rate of change of altitude occurs at the instants of rising and setting. Bodies whose declinations are of the same name as, but of greater magnitude than, that of the observer's latitude, change their altitudes most rapidly when they are at their limiting azimuths.

Formula (2) indicates that when the azimuth of a body is 0° or 180°; that is to say, when the body is at meridian passage at either upper or lower transit, its rate of change of altitude is zero.

Formula (2) may be derived directly by using the differential calculus as follows. The spherical cosine formula applied to the astronomical triangle PZX is:

$$\cos z = \cos p \sin \phi + \sin p \cos \phi \cos h$$

Differentiating with respect to h, we have:

$$\sin z \frac{dz}{dh} = \sin p \cos \phi \sin h$$

From the spherical sine formula:

$$\sin Z = \frac{\sin p \sin h}{\sin z}$$

Therefore, by substitution we have:

$$\frac{dz}{dh} = \cos \phi \sin Z$$

Now dz/dh is the rate of change of zenith distance, the magnitude of which is equivalent to the rate of change of altitude. If the rate is expressed in minutes of arc per minute of time, we have, as before:

Rate of change of altitude $= (15 \cos \phi \sin Z)'$ per minute

If the changing altitude of a fixed celestial body be graphed against time, the resulting curve will be symmetrical about an ordinate corresponding to the time of meridian passage. This ordinate will represent the meridian altitude which corresponds to the maximum altitude of the body for the day. This will apply only to fixed celestial bodies observed by a stationary observer. For a stationary observer the curve of changing altitude against time applicable to the Sun, and even more particularly to the Moon, is not generally a curve symmetrical about the ordinate representing the meridian altitude. This results from the changing declination of the body.

The rate of change of the Sun's declination varies as the cosine of its declination. A change in the Sun's declination amounting to $(2 \times 23\frac{1}{2})°$, that is $47°$, takes place in the six-monthly period between solstices. At an equinox, when the Sun's declination is zero, his rate of change of declination is greatest, being about 1' per hour: at a solstice, when the Sun's declination is maximum, his rate of change of declination is zero. It follows that, to a stationary observer in a high northerly latitude, the Sun continues to increase his altitude after the time of meridian passage during the period when he changes his declination towards the north, that is to say, during the period between the Winter and Summer solstices. On March 21st, when the Sun's rate of change of declination towards the north is greatest, the interval between meridian passage and the instant when he attains his maximum altitude is greatest.

During the period between Summer and Winter solstices, when the Sun's declination increases towards the south, he will attain his maximum altitude to an observer in a high northerly latitude before he is at meridian passage. The interval between the times of meridian and maximum altitudes during this period will be greatest on the day of the Autumnal equinox, that is on September 23rd.

The rate of change of the Moon's declination, like that of the Sun's, varies as the cosine of her declination. The Moon's maximum declination may be as much as $28\frac{3}{4}°$ N. or S. The plane of the Moon's orbit is inclined at an angle of $5\frac{1}{4}°$ to the plane of the ecliptic. The points of intersection of these planes on the celestial sphere—points known as the *nodes*—swing with retrograde motion around the ecliptic once in 18·6 years. The node at which the Moon lies when changing her declination from south to north is called the *ascending node*: the other node is called the *descending node*.

Depending upon the position of the ascending node relative to the First Point of Aries, the Moon's declination during any sidereal period, may be anything between $(23\frac{1}{2} - 5\frac{1}{4})°$ N. or S. and $(23\frac{1}{2} + 5\frac{1}{4})°$ N. or S.

When the Moon's declination is changing towards the north she will continue to increase her altitude to a stationary observer in a high northerly latitude, after she has crossed the observer's upper meridian. Her maximum altitude in these circumstances is attained after the time of her meridian passage. When the Moon's declination is changing towards the south her maximum altitude, to a stationary observer in the northern hemisphere, will occur before the time of her meridian passage. For a stationary observer in a high southerly latitude, the reverse will apply.

If at the time of meridian passage, the declination of the Sun (or Moon) is increasing at the rate of x' per minute, maximum daily altitude to a stationary observer will occur when the rate of change of altitude due to the Earth's rotation is decreasing at the rate of x' per minute.

Not only does the rate of change of declination affect the rate of change of the altitude of a celestial body: any movement of the observer will also affect the curve of changing altitude against time. Let us deal first with the effect of movement of an observer towards the east or west. Any such movement results in a corresponding movement of the observer's celestial meridian.

Because the Earth spins towards the east, celestial bodies tend to revolve around the Earth towards the west at the rate of 15° per hour. Any movement of an observer towards the east, therefore, tends to increase this rate. Conversely, any movement of the observer towards the west tends to decrease it. It follows that when an observer is moving eastwards over the Earth's surface, the rate of change of altitude of a celestial body is greater than it is for a stationary observer. When an observer is moving towards the west, a body's rate of change of altitude is less than it is for a stationary observer.

Easterly or westerly movement of an observer has no effect on the symmetry of the curve of changing altitude against time. It merely accelerates or decelerates the rate of change of altitude. Northerly or southerly motion of an observer over the Earth's surface does, however, affect the symmetry of the curve.

The effect of meridianal movement of an observer is the same as that of changing declination of a heavenly body. When an observer is moving such that his zenith is approaching a fixed celestial object at meridian passage, the object will attain its maximum altitude after the time of meridian passage. Conversely when an observer's zenith and a fixed celestial object at meridian passage are separating the object will attain its maximum altitude before the time of its meridian passage.

To those navigators who pin their faith in the latitude by

meridian altitude of the Sun, it should be important that the observation be made at the time of meridian passage. The normal practice of waiting until the Sun dips and using the maximum altitude in place of the meridian altitude, may lead to considerable error in the latitude of a ship travelling at great speed northerly or southerly.

It will be interesting, in view of the above remarks, to investigate a formula for finding the interval between the instants of meridian and maximum altitudes.

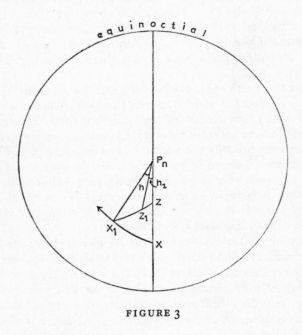

FIGURE 3

Fig. 3 illustrates the northern celestial hemisphere in the plane of the equinoctial. P is the elevated celestial pole and Z is the zenith of an observer at the instant a celestial body is at upper meridian passage at X. Z_1 is the observer's zenith at the instant the body is at its maximum altitude at X_1.

Applying the spherical sine formula to the spherical triangle PZ_1X_1, in which the angle P (denoted by h) is the hour angle of the body when it is at maximum altitude, we have:

$$\frac{\sin Z_1}{\sin PX_1} = \frac{\sin P}{\sin Z_1X_1}$$

from which:

$$\sin Z_1 = \frac{\sin h \cos d}{\sin Z_1X_1} \tag{1}$$

Let x represent the rate of change in the ship's longitude towards the west. Let y represent her rate of change of latitude combined with the rate of change in the declination of the body in the same interval, expressed in minutes of arc per hour.

If the ship's change of latitude is in the same sense as the change in the object's declination, y is the difference between the changes in the observer's latitude and the object's declination: if they are opposite in sense, y is the sum of these.

We have seen that for a stationary observer the rate of change of altitude of the Sun is:

$$(15 \cos \phi \sin Z)' \text{ per minute}$$

or

$$(900 \cos \phi \sin Z)' \text{ per hour}$$

Taking into account the change in the ship's longitude (x' per hour towards the west), this formula becomes:

$$\text{Rate} = [(900 - x) \cos \phi \sin Z]' \text{ per hour}$$

The Sun reaches its maximum altitude when the ship's rate of northerly or southerly motion is equivalent to this. It follows that when the body is at maximum altitude:

$$y = (900 - x) \cos \phi \sin Z$$

from which:

$$\sin Z = \frac{y \sec \phi}{900 - x}$$

Substituting this value for $\sin Z_1$ in equation (1) we get:

$$\frac{\sin h \cos d}{\sin Z_1X_1} = \frac{y \sec \phi}{900 - x}$$

from which:

$$\sin h = \frac{y \sec \phi \sin Z_1X_1}{\cos d (900 - x)}$$

But Z_1X_1 is equivalent to $(\phi \pm d)$, therefore:

$$\sin h = \frac{y \sec \phi \sin (\phi \pm d)}{\cos d (900 - x)}$$

$$= \frac{y \sec \phi (\sin \phi \cos d \pm \cos \phi \sin d)}{\cos d (900 - x)}$$

$$= \frac{y \sec \phi (\sin \phi \pm \cos \phi \tan d)}{900 - x}$$

$$= \frac{y (\tan \phi \pm \tan d)}{900 - x}$$

Now h is small when the celestial body is near meridian passage, and in these circumstances we may assume the equivalence of $\sin h$ and h radians without introducing material error. Thus:

$$h^\circ = \frac{y(\tan \phi \pm \tan d)}{900 - x}$$

and

$$h' = \frac{3438y(\tan \phi \pm \tan d)}{900 - x}$$

and

$$h \text{ secs} = \frac{4 \times 3438y(\tan \phi \pm \tan d)}{900 - x}$$

Now $\dfrac{1}{900 - x} = \dfrac{1}{900\left(1 - \dfrac{x}{900}\right)} = \dfrac{1}{900}\left(1 - \dfrac{x}{900}\right)^{-1}$

Since $x/900$ is a very small quantity we may assume that:

$$\frac{1}{900 - x} = \frac{1}{900}\left(1 + \frac{x}{900}\right)$$

so that:

$$h \text{ secs} = 4 \times 3438y(\tan \phi \pm \tan d)\left(1 + \frac{x}{900}\right)/900$$

That is:

$$h \text{ secs} = 15 \cdot 28 y (\tan \phi \pm \tan d)\left(1 + \frac{x}{900}\right) \qquad (2)$$

Referring to Fig. 3, let the interval between the instants of meridian and maximum altitudes be H secs. Then:

$$H = ZP_n Z_1 + Z_1 P_n X_1$$

i.e. $$H = h_1 + h$$

This applies to the case when maximum altitude occurs after the time of meridian passage. When maximum altitude occurs before the time of meridian passage:

$$H = h_1 - h$$

Let the interval between the times of meridian and maximum altitudes be H where:

$$H = h + h_1$$

Now h_1 represents the change in the ship's longitude between the instants of meridian passage and maximum altitude. That is to say h_1 is the change of longitude in the time interval H.

$$\frac{h_1}{H} = \frac{\text{Rate at which ship is changing her long.}}{\text{Rate at which Sun is changing his H.A.}} = \frac{x}{900}$$

i.e. $$h_1 = \frac{Hx}{900}$$

Now
therefore: $$H = h_1 + h$$

$$H = \frac{Hx}{900} + h$$

and $$H = h / \left(1 - \frac{x}{900}\right)$$

or $$h = H\left(1 - \frac{x}{900}\right)$$

therefore:

$$H = h\left(1 - \frac{x}{900}\right)^{-1}$$

Since $x/900$ is a very small quantity we may assume that:

$$H = h\left(1 + \frac{x}{900}\right)$$

or

$$h = H/\left(1 + \frac{x}{900}\right)$$

Substituting this value for h in equation (2) we have:

$$H \text{ secs} = 15\cdot28y(\tan \phi \pm \tan d)\left(1 + \frac{x}{900}\right)^2$$

And again since $x/900$ is a very small quantity this reduces to:

$$H \text{ secs} = 15\cdot28y(\tan \phi \pm \tan d)\left(1 + \frac{x}{450}\right)$$

If the ship changes her longitude towards the east during the interval between meridian passage and maximum altitude the formula becomes:

$$H \text{ secs} = 15\cdot28y(\tan \phi \pm \tan d)\left(1 - \frac{x}{450}\right)$$

CHAPTER VI

Errors in Astronomical Navigation

In this chapter we shall discuss the principal errors that occur in the processes of practical astronomical navigation.

Nautical astronomy is not an exact science. The measurements made by the navigator when using sextant or chronometer, the quantities extracted from the *Nautical Almanac* and from nautical tables, and the computational processes employed when reducing sights, are all liable to error. An intelligent nautical astronomer should aim to understand the nature of the several errors which may influence the degree of accuracy of his observed positions.

The required degree of accuracy of any navigational result varies with the use to which the result is to be put. If, for example, a navigator wishes to keep his ship in a channel, then assuming that he knows that his ship is somewhere near the middle of the channel, the required degree of accuracy of his fixes should be related to the channel width. It would be pointless to try to fix his ship to the nearest cable if the width of the channel were 10 miles. On the other hand, an accuracy to the nearest cable would be insufficient if the channel width were only a cable. Again, a compass error worked out to the nearest minute of arc (often done in class- and examination-rooms) serves little better than one worked out to the nearest quarter or even half degree, bearing in mind the relative coarseness with which compass bearings are measured.

Safe navigation requires the navigator to fix his ship and set her courses within certain safe limits. Any combination of errors within these safe limits will not endanger the ship. The wider are the safe limits prescribed by the navigator the smaller will be the required degree of accuracy of the quantities and processes involved in producing a navigational result.

Before discussing navigational errors a few remarks on arithmetic and its processes will be relevant to our purpose. By arithmetic we mean the mathematics of computation in which numbers are used. The fractional numbers used in most navigational processes are expressed in decimals. It is important to realize that nothing is to be gained, and considerable time and effort may be wasted, when a number is expressed with a precision greater than that justifiable. The *precision* of a decimal quantity is indicated by the number of digits to the right of the decimal point. Consider the quantity the magnitude of which is 13·64 precisely. The quantity may be described as 13·6 or 14 to mean respectively that its magnitude lies between 13·55 and 13·65 or between 13·5 and 14·5. We say, therefore, that if the magnitude of the quantity is described as 13·6 the description is more precise than by giving it as 14, etc.

In expressing a number which is a multiple of 10, 100, 1000, etc., there is sometimes a doubt as to its precision. If, for example, the tonnage of a ship is described as 8000 tons, the tonnage may be taken to mean between 7500 and 8500; or between 7900 and 8100; or between 7990 and 8010 tons.

A common way of expressing the degree of precision of a numerical quantity is to state the number of *significant figures*. Significant figures in a number, as the name implies, are those that occupy places which indicate their significance. For example, in the number 82·07 there are four significant figures, for we know that the symbols represent eight tens, two units, nought tenths, and seven hundredths, respectively.

In determining the number of significant figures in a number, caution is necessary in respect of zeros. Zeros interspersed between digits are always significant figures (as in the example above) but zeros written to the right of a digit in a whole number present difficulty in determining which figures are significant. In the above example, we could say that the tonnage of the ship is 8000 tons to four, three or two, significant figures, according to whether we wish it to mean 8000 tons precisely; or between 7990 and 8010; or between 7900 and 8100 tons, respectively.

It is important to bear in mind that in any arithmetical computation the result can never be more accurate than the least precise value used. An example will illustrate this. Suppose we wish to add 10, 6·4 and 5·35. From the above remarks 10, as written, may mean 'between 9·5 and 10·5'; 6·4 may mean 'between 6·35 and 6·45'; and 5·35 may mean 'between 5·345 and 5·355'. It follows that the result of the addition may lie between (9·5 + 6·35 + 5·345), i.e. 21·195; and (10·5 + 6·45 + 5·355), i.e. 22·305. The sum of the three numbers is 21·75 precisely only if the numbers are 10, 6·4 and 5·35 precisely in each case. If the three numbers are not precise the result 21·75 may give the computer a false indication of accuracy.

It is useful to distinguish between the terms *accuracy* and *precision*. In many, if not all, navigational processes we deal with measurements of quantities which are continuous as opposed to those which are composed of discrete and separate elements. The quantities measured by a navigator when using sextant or chronometer are called by arithmeticians *approximate numbers*.

An approximate number is incorrect because of the *error* that exists between it and its true value. Let us suppose that the length of the rod illustrated in Fig. 1 is to be measured with each of three rulers, labelled in the figure A, B and C, divided to inches, half-inches and quarter-inches, respectively.

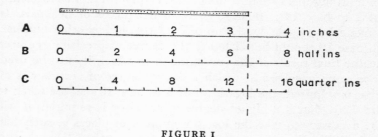

FIGURE I

Using ruler A we find that the rod is 3 inches long to the nearest inch. Using B we find it to be (7 × ½), i.e. 3½ inches to the nearest half-inch, and using C we find the length to be (13 × ¼), i.e. 3¼ inches to the nearest quarter-inch. In each case the length of the rod is expressed in terms of the nearest exact unit of measurement of the ruler. The result, therefore, is accurate to within a half unit of the measurement given. The accuracy of the result using A is within $\pm\frac{1}{2}$ inch of the true value. That using ruler B is within $\pm\frac{1}{4}$ inch of the true value; and that using ruler C is within $\pm\frac{1}{8}$ inch of the true value.

The term precision is used to denote the degree of accuracy of an approximate number. The smaller is the degree of accuracy the more precise is the measurement. Precision and accuracy are the concern of the navigator, not only when using navigational instruments, but also when using logarithms and trigonometrical functions.

Most logarithms and trigonometrical functions are irrational numbers and, with few exceptions, entries in tables of logarithms and trigonometrical functions are but approximations of exact values.

The logarithm of a number to a given base (10 in the case of common logarithms) is the power to which the base must be raised to give the number. Thus, the logarithm of N to base B is L, that is:

$$\log_B N = L$$

or $$B^L = N$$

The logarithm of 100 to base 10 is 2 precisely. The logarithms of most numbers are unending decimal quantities. The logarithm of 2 to base 10 is 0·3011385557 to ten places of decimals. In practical work, the logarithm of 2 (or that of any other number) to base 10 should be related to the degree of precision required in the final result of the computation in which it is to be used. For some purposes in which a coarse degree of accuracy is all that is required, the logarithm of 2 to base 10 may be taken to be 0·301, but if a higher degree of accuracy is required, it will be necessary to take the log to a number of places greater than three. The last figure of an incommensurable logarithm is called a *rounded figure*.

By *rounding off* we mean expressing a decimal quantity to a degree of accuracy less precise than the accuracy of the given quantity. If the digit in the extra (or unwanted) place is 4 or less it is ignored when rounding off. If the digit in the extra place is 5 or more, the digit in the last desired place is increased by 1. Thus the number 8·656 is rounded off to two decimal places to 8·66, and to one decimal place as 8·7.

In logarithmic tables the final figure is rounded off and may be in excess or defect of half a unit. For example, the log of 2·25 is 0·3521825 to seven figures. To five figures it is 0·35218, so that the relative error in the fifth figure is 0·25 defect. To four figures it is 0·3522 so that the relative error in the fourth figure is 0·0175. It is interesting and important to appreciate the error introduced into results obtained by logarithmic computation.

In extracting from a table a logarithm or trigonometrical function, there may be an error of 0·5 in the last place. If, therefore, two logarithms are combined by addition or subtraction, the combined error may be anything between +1 and −1 in the last place.

In lifting from a table a number corresponding to a given logarithm there are two sources of error. First the ±0·5 in the tabulated log and second the ±0·5 in the calculated log. At worst, therefore, there is an error of ±1 in the last place of logarithms to cause an error in the result. It is interesting to work out the percentage errors that apply when using logarithms to various places of decimals. Five-figure logarithms are amply sufficient for astronomical navigational purposes. The maximum percentage error using five-figure logarithms is about 0·002. Four-figure logarithms used for nautical astronomical problems generally give an accuracy to within one minute of arc.

There are cases in which satisfactory results cannot be expected regardless of the number of figures in the logarithms. Examination of the table of log sines of large angles will exemplify this. The computer should recognize these so-called ill-conditioned cases, in which a very small error in the data is magnified greatly in the result. Accordingly he should distrust the result when these cases are forced upon him. Now let us turn our attention to a discussion on errors.

The term error applies to the difference between a correct and a corresponding incorrect value arising from imperfections or the instruments or methods used in obtaining a result. Errors are to be distinguished from what statisticians euphemistically call *blunders*. Blunders are due largely to carelessness and are commonly called mistakes. Before dealing with errors a few words on blunders and how to avoid them will be made.

Blunders in nautical astronomy may arise from careless reading of sextant or chronometer. Sextant readings are commonly in error by 10°, 1° or 10′. After having read the sextant, it is advisable to leave the index bar in its original position until after the sight has been worked. By so doing the sextant reading may be checked if necessary. Chronometer times are commonly in error by an hour, or a multiple of five minutes, or a minute. In order to check chronometer times it is a good practice to have the chartroom clock time recorded for the instant of observation.

The fact that the chronometer dial registers only 12, and not 24 hours often leads to an error of 12 hours in the G.M.T. No self-respecting nautical astronomer makes this mistake in practice; for, knowing the approximate local time and the ship's longitude, it is a simple matter to ascertain an approximate G.M.T. which may be used in deciding whether the chronometer time is equal to, or 12 hours different from, the approximate G.M.T.

Careless use of the *Nautical Almanac* often gives rise to blunders. Using the wrong month and the correct day of the week, or the correct month and the wrong day of the week, are all too common careless mistakes made by navigators. It is customary, and good practice, to cross out each day's date in the *Nautical Almanac* as soon as convenient after the Greenwich date changes.

Many navigators solve their sights by long methods which often lead to blunders in arithmetic. A neat layout of the working of the sight is essential if arithmetical mistakes are to be kept to a minimum. Addition of groups of figures should be checked upwards or downwards in the reverse direction to that used in the first attempt. Subtractions should be checked by adding the difference to the lesser value. Care is necessary when converting from one unit to another. A common mistake, for example, is to count 100 minutes of arc, instead of 60, to a degree, when adding or subtracting angular measures.

Mistakes due to lack of facility in simple arithmetical processes lead to considerable frustration and lack of confidence. Navigators who are prone to making arithmetical mistakes should, perhaps, use inspection tables or short-method tables, instead of long methods, for solving their sights.

Amongst the more common blunders that arise when using nautical tables is that due to reading from the wrong end of the table. For example, some log-trig function tables and traverse tables are downward reading for angles between 0° and 45°, and upward reading for angles between 45° and 90°. Great care should be exercised when using these tables.

Other common blunders due to carelessness arise from applying corrections, such as altitude corrections, index error of sextant, etc., in the wrong direction or sense.

When plotting position lines, blunders resulting from: applying the intercept the wrong way, drawing a position line in the direction of the azimuth instead of at right angles to it, and using the wrong scale in marking off intercepts, are not uncommon.

It appears that blunders in arithmetic occur most frequently when performing relatively simple computational tasks which do not require a large measure of intelligent concentration. Boredom, overconfidence, or unnecessary hurry, often lead to blunders. These factors clearly are related to temperament.

Errors in navigation may be *systematic* or *random*. A systematic error is one which follows a set pattern. If the pattern is understood a systematic error may be predicted. A common systematic error in nautical astronomy is index error of the sextant. Index error, if it is measured immediately before or after making an altitude observation, may be allowed for by applying a correction of equal magnitude but opposite in sense to the index error, to the sextant altitude. Index error is an example of a constant systematic error.

Personal error, due to habitually over- or underestimating an observed altitude, is another example of a constant systematic error which may be dealt with in the same way as sextant index error.

Systematic error may result from not applying the effect of a current in working out an estimated position. The error, in this case, is proportional to the time during which the current acted.

Another type of systematic error may arise through faulty interpolation. Interpolation is the process of finding the value of an element which falls between two given values. Interpolation is a very commonly used process in navigation. Values of the elements given in the *Nautical Almanac*, for example, apply to particular values of G.M.T. If the given G.M.T. (for which an element is required) is different from a tabulated G.M.T., it will be necessary to interpolate between the next higher and next lower G.M.T. to the given G.M.T.

The process of finding an element the value of which depends upon more than one argument; such as, for example, the process of finding the amplitude of a heavenly body from a table requiring double entry of latitude and declination, is called *double interpolation*. When three arguments are required for finding an element; as, for example, in finding the azimuth of a celestial body from Davis's or Burdwood's tables (in which azimuths are tabulated against latitude, declination and local hour angle), the process is called *triple interpolation*.

The smaller is the interval between tabulated arguments the smaller is the need for careful interpolation. If the interval is large great care is necessary when interpolating.

In practice many tables in which it may be necessary to interpolate are designed so that the intervals between tabulated arguments are sufficiently small to assume that the value of the element changes directly as that of the argument. In this case, a graph representing values of the element against an argument may be considered to be a straight line. Interpolation in this case is called *linear interpolation*.

Linear interpolation, when only one argument is involved, is accomplished by simple proportion. Suppose the tabulated values of declination are given at integral hourly intervals as follows:

Time	Declination
	° ′
0200	12 16
0300	12 22
0400	12 28

In finding the declination for 0220, a correction is applied to the declination for 0200. The correction is found from the relationship:

$$\frac{c}{d} = \frac{t}{T}$$

where c = required correction to the tabulated declination,

d = difference between successive tabulated declinations between which it is required to interpolate,

t = difference between the given and tabulated times,

and T = difference between successive tabulated times.

The correction in the example is $(20 \times 6)/60$, i.e. $2'$, so that the required declination is $12° 16' + 2'$, i.e. $12° 18'$. The correction in this case is additive because the declination increases with time.

Linear interpolation in practice is usually performed by mental arithmetic.

The practical method for triple interpolation is first to extract the element corresponding to the tabulated arguments equal to or less than those given, and then to apply three corrections to this element, each correction being obtained by single linear interpolation. An example will clarify this.

EXAMPLE: Find the true azimuth of the Sun whose L.H.A. is 3 hr 05 min W. and whose declination is $20° 18'$ N. to an observer in latitude $49° 24'$ N. Use Burdwood's Azimuth Tables.

In Burdwood's tables azimuths are tabulated for each integral degree of latitude and declination and for every degree of local hour angle. To interpolate we proceed as follows:

	Lat °	Dec °	H.A. h m		
Azimuth for	49 N.	20 N.	3 04 =		111·7°
Azimuth for	49 N.	20 N.	3 08 =	110·7°	
		correction for H.A. = $\frac{1}{4} \times 1·0$	= −0·3		
Azimuth for	49 N.	21 N.	3 04 =	110·6°	
		correction for dec = $\frac{3}{10} \times 1·1$	= −0·3		
Azimuth for	50 N.	20 N.	3 04 =	112·5°	
		correction for lat = $\frac{4}{10} \times 0·8$	= + 0·3		
Azimuth for	49 24'	20 18'	3 05 =		111·4

Required azimuth = N. $111\frac{1}{2}°$ W.

Single interpolation may be obviated by arranging a single argument table so that the table is entered within limiting values of the argument. If the argument happens to be a limiting value the entry is said to be critical, in which case a choice of two possible values of the required element has to be made. Some of the altitude correction tables in the *Nautical Almanac* are critical tables. In these tables the correct element for a critical entry is the upper of the two possible values.

When the rate of change of a tabulated element is not proportional to that of the argument, linear interpolation will result in error in the extracted element. Examples of non-linear rates of change in navigation are the rates of change of the hour angle and declination of the Moon and planets. When interpolating for these quantities, using the linear interpolation tables provided in the *Nautical Almanac*, it is necessary to apply a secondary correction, called the *v* or *d* correction, to the main correction lifted from the interpolation tables.

In some cases a table may be 'extended' by a process known as *extrapolation*. By extrapolating we mean obtaining an element from an argument lying outside the limits of the table. Extrapolation is reliable only when the rate of change in the elements outside the tabular limits is linear.

We now come to a brief discussion on a type of error known as a random error. In contrast to systematic errors random errors are those which cannot be predicted. These include errors that are inherent in practical work. In navigation random errors are seldom very big. In measuring an altitude with a sextant the accuracy of the measured result is influenced by several factors, which include: indistinct horizon, abnormal refraction, changing height of eye due to rolling, pitching or heaving of the ship. Moreover the limit of accuracy with which the sextant can be read may also result in error. All of these are examples of random or chance errors.

Random errors are governed by the mathematical laws of probability. The term probability may be defined as the proportional frequency of occasions on which some stated event takes place. If for example a series of sextant altitude observations is made, and the observations are liable to a random, but not a systematic, error, the probability of the result of any given observation being greater or less than the corresponding true value is expressed as 0·5. This follows because the probability or chance of a particular observation producing too high a result is equal to that of its producing too low a result.

If a large number of observations affected by a random error were made and the results plotted as a frequency curve in which error is plotted against the probability of it happening, the result would tend to be a curve known as a *Normal* or *Gaussian curve of errors*. Fig. 2 illustrates such a curve.

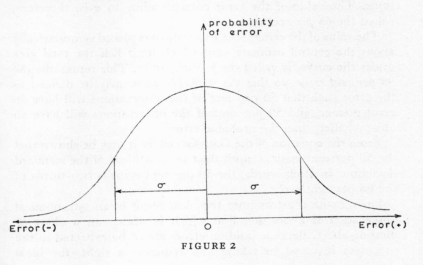

FIGURE 2

The Gaussian curve is a bell-shaped curve symmetrical about an ordinate representing the proportional frequency of observations yielding the correct value. The ordinates of points on the curve to the right of the central ordinate represent the proportionate frequency of observations yielding too high a result, and those of points to the left represent the proportionate frequency of observations yielding too low a result.

The curve illustrates that the possibility of a random error occurring falls off as the size of the error increases.

To express the 'average' error in a series of observations which produce a Gaussian curve statisticians use a quantity called *standard deviation*. This is found by squaring each error, dividing the sum by the number of observations, and taking the square root of the quotient. The result is sometimes called the root mean square (R.M.S.) and is denoted by the Greek letter sigma (σ). It is of interest to note that the abscissae at the points of inflexion on the Gaussian curve are each equivalent to the standard deviation.

If the total area under a Gaussian curve represents 100 per cent of the observations, the area between the ordinates at the two points of inflexion represents 68 per cent of the observations. The value of the error corresponding to σ is, therefore, called the *68 per cent error*.

The value of the error at the two ordinates placed symmetrically about the central ordinate, and which limit half the total area under the curve, is called the *probable error*. This represents the *50 per cent error*, so that the probable error may be defined as the error such that 50 per cent of the observations will have an error greater, and 50 per cent of the observations will have an error smaller, than the probable error.

From the equation of the Gaussian curve it may be shown that the 50 per cent error is equivalent to two-thirds of the standard deviation. In other words, the 50 per cent error is two-thirds of the 68 per cent error.

For a nautical astronomer the ideal result of an astronomical observation is a position line which is drawn on a chart or plotting sheet. Because random errors are to be expected in the processes involved in taking and reducing a sight, the ideal

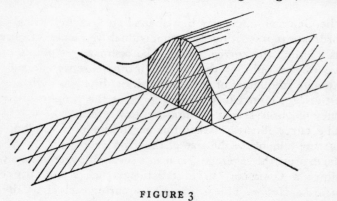

FIGURE 3

result, viz. a true line of position on the chart, is not to be expected. The practical result of an astronomical observation is a position band on the chart the width of which may be regarded as being proportional to the probable, that is the 50 per cent error. The band may be considered to be the projection on the chart of part of a ridge the shape of which, at right angles to the band, corresponds to the Gaussian curve of errors. This is illustrated in Fig. 3.

The band of 50 per cent error may conveniently be regarded as being formed by a series of parallel lines symmetrically disposed about the mid-line of the band. These lines represent the projections of ordinates bounding equal areas under the Gaussian curve as illustrated in Fig. 4.

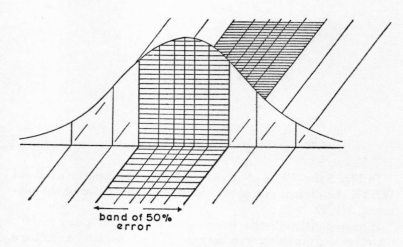

band of 50%
error

FIGURE 4

The distance between any pair of adjacent lines in the band of error illustrated in Fig. 4 is such that the chance of the ship lying between the pair is equal.

A fix in the simplest case is obtained by crossing two position lines. In the ideal case, in which the position lines are free from error, the required fix is the point of intersection of the two position lines. If, on the other hand, two position lines are affected by random error, the 50 per cent bands of error will intersect to form a diamond.

It may at first be thought that there is a 50 per cent chance of the ship's position falling within the diamond formed by the intersection of the 50 per cent bands of error. This is not the case. Statistical analysis reveals that the 50 per cent probability area is an ellipse which fits into a diamond having dimensions 1¾ times those of the diamond formed by the intersecting 50 per cent position bands.

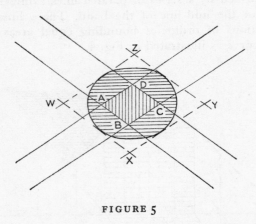

FIGURE 5

In Fig. 5 the ellipse of 50 per cent error fits into the diamond WXYZ the dimensions of which are 1¼ times those of the diamond ABCD.

If two position bands have equal width and cross at 90°, the ellipse of error becomes a circle. If the position bands have equal widths but cross at any angle other than 90°, the major axis of the ellipse of error bisects the acute angle between the bands. If the position bands have unequal widths and cross at any acute angle, the major axis of the ellipse of error lies nearer to the narrower than to the broader band.

A navigator, in recognizing the possibility of errors affecting his position lines, at once realizes that the concept (common amongst navigators) that information used in computing a ship's position is perfectly reliable, is false and unrealistic. A ship's position obtained as a result of a systematic evaluation of the information used in getting it may be described as a probable position, because recognition of probable error has been made.

Probable error in a position line is estimated: there is no other practical way of evaluating it. The estimation of probable error in nautical astronomy is related to a man's skill and experience as a navigator. Skill, in this connection, is related to the understanding of the nature of the errors which may affect navigational observations and processes. Without this understanding errors cannot be handled systematically and intelligently. The practical treatment of navigational errors will be discussed in Part IV, Chapter V.

PART III

The Instruments
of Nautical Astronomy

CHAPTER I

The Sextant

The sextant is the distinctive instrument of the nautical astronomer. In astronomical navigation the sextant is used for measuring altitudes of celestial bodies, and it is with this aspect of its use that this chapter will primarily be concerned.

The modern sextant dates from the middle of the 18th century. It has evolved from the reflecting quadrant the invention of which is usually attributed to John Hadley, a prominent English philosopher of the period. The sextant is a portable instrument which is ideally suitable for measuring altitudes from the unsteady platform of a rolling, pitching or heaving ship. The name given to the nautical instrument is derived from the fact that the graduated arc against which observed altitudes are measured is a sixth part of a circle. The sextant is described as an instrument of double reflection, and it may be used for measuring angles up to about a third of a circle.

Although sextants vary widely in details of design, and with the auxiliary equipment provided by the manufacturer, the principal constructional features are common to all sextants. These features are illustrated in Fig. 1.

The *frame* labelled A in Fig. 1 is usually of brass or aluminium. The *limb* B is graduated in degrees from about $-5°$ to $+125°$. Pivoted to the frame at the centre of curvature of the limb is the *index bar* C. The index bar is so-named because it carries the pointer or *index* by means of which the angle corresponding to the measured altitude is read against the graduated scale on the limb. Mounted on the index bar is the cell which accommodates the *index mirror* D.

The frame is provided with a collar E which houses the *sextant telescope* F. In line with the axis of the telescope is the *horizon glass* G which is accommodated in a cell fitted to the frame. The horizon glass is generally half-silvered, that half adjacent to the frame being silvered and the other half plain.

Mounted on the index bar tangentially to the limb is a screw which engages in teeth cut in the limb. The index bar may be moved from one end of the arc to the other by turning the *tangent screw*. For this reason the screw is called an *endless* tangent screw. The index bar is fitted with a spring-actuated

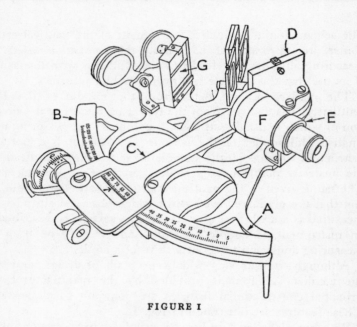

FIGURE I

clamp by means of which the tangent screw may be disengaged. This facilitates setting the index bar to any desired position on the arc. One end of the tangent screw carries the *micrometer drum*. One complete turn of the drum causes the index to move exactly one degree of arc across the scale graduated on the limb. The circumferential surface of the micrometer drum is graduated in minutes of arc. Adjacent to the drum and fixed to the index bar is a *Vernier scale*, by means of which altitudes may be measured to a fraction—usually one-tenth—of a minute of arc.

To reduce the brilliance of the rays of light entering the observer's eye from the observed object—especially the Sun— and the horizon, *tinted shades* are fitted. Those for use with the reflected ray from the index mirror are called the *index shades*; and those for use with the direct ray from the horizon are called the *horizon shades*.

The sextant frame is fitted with a handle of wood or plastic. This provides a convenient place for accommodating a dry electric cell which is connected to a bulb fitted to the index bar for use when reading star altitudes.

Before the introduction of the micrometer drum sextant, the so-called *Vernier sextant* was in general use. This type of sextant, in which the arc scale is usually cut to 10' intervals, is provided with a Vernier scale fitted on the index bar, by means of which altitudes may be read to an accuracy of 10" of arc. Vernier sextants are now obsolescent.

Of academic interest to the navigator is the optical principle of the sextant which is based on the elementary laws of light, viz.:

1. When a ray of light is reflected from a mirror the incident and reflected rays and the perpendicular or normal to the reflecting surface at the point of reflection are co-planar.
2. The angle which the incident ray makes with the normal is equal to the angle which the reflected ray makes with the normal. In other words, the angles of incidence and reflection are equal.

Arising from these two laws, the optical principle of the sextant is:

When a ray of light is successively reflected from two mirrors the reflecting surfaces of which are perpendicular to a common plane, the angle between the reflecting surfaces is half the angle between the first incident and the final reflected rays.

Fig. 2 illustrates the optical principle of the sextant.

In Fig. 2 XABC is the zig-zag ray which is doubly reflected at A and B. N_1AN_2 and BN_2 are the normals to the mirrors at A and B respectively.

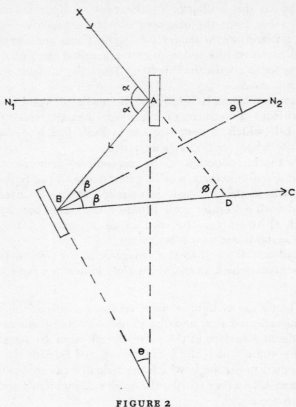

FIGURE 2

From the laws of light:

$$XAN_1 = N_1AB = \alpha$$
$$ABN_2 = N_2BC = \beta$$

Let the angle between the first incident ray XA and the final reflected ray BC be ϕ.

Now the angle between the reflecting surfaces is equal to the angle between the normals to the reflecting surfaces at the points of reflection. Let this angle be θ.

In triangle BAN_2: $\theta = (\alpha - \beta)$

In triangle ABD: $\qquad \phi = 2\alpha - 2\beta$

$$= 2(\alpha - \beta)$$

therefore: $\qquad \phi = 2\theta$

It follows that the angle between the mirrors is half the angle between the first incident and the final reflected rays. For this reason the 60° arc of the sextant is graduated to 120 divisions, each division representing 1° of measured altitude.

The sextant is in correct adjustment when the index on the index bar is at zero on the graduated arc, with the index mirror and the horizon glass parallel to one another, and both mirrors perpendicular to the plane on which the arc lies. Moreover, the axis of the sextant telescope should be parallel to the plane of the sextant. If a sextant is not in perfect adjustment the error in a measured altitude may be considered to be a combination of several component errors. These components are of two types, known respectively as *adjustable* and *non-adjustable errors*. Let us deal first with the adjustable errors.

There are four adjustable errors which may be eliminated by making the *first, second, third* and *fourth adjustments*.

The first adjustment is made in order to set the index mirror perpendicular to the plane on which the arc lies. The cell in which the index mirror is housed is fitted with a spring which bears against the back of the mirror. A screw adjustment is made to cause the spring to bring and keep the plane of the mirror perpendicular to the plane of the arc.

If the index mirror is not set correctly the sextant possesses *error of perpendicularity*. It is an easy matter to test the sextant to ascertain if the cause of this error exists. The instrument is held horizontally face upwards and with the arc away from the observer. The index bar is set near the middle of the arc. By observing the reflection of the arc in the index mirror and comparing its alignment with the true image of the arc to the right of the mirror, the observer readily can see whether or not the index mirror is properly set. If the true and reflected images of the arc are not in the same straight line, as illustrated in Fig. 3(a), it will be necessary to make the first adjustment.

The second adjustment is made in order to set the horizon

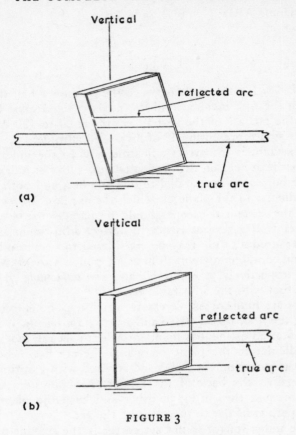

FIGURE 3

glass perpendicular to the plane of the arc. The adjustment is made by means of a screw at the top of the back of the cell which houses the horizon glass.

If the horizon glass is not set correctly the sextant possesses *side error*.

To test the perpendicularity of the horizon glass the sextant is held with the arc in the vertical plane and with the index set near the zero on the arc. A star is observed so that the position of the reflected image of the star may be compared with the star's direct image as seen through the unsilvered part of the horizon glass. If one of the images lies to the side of the other, the sextant possesses side error, and it will be necessary to make the second adjustment.

The cause of side error may be detected in the daytime by observing a vertical edge (such as that of a mast or building) with the sextant arc lying in the vertical plane; or by observing a horizontal edge (such as the horizon) with the sextant arc lying in the horizontal plane. If the line through the reflection of the observed edge is not co-linear with the direct image observed through the unsilvered part of the horizon glass, the sextant possesses side error.

The second adjustment is normally made simultaneously with the third adjustment. The third adjustment is necessary if the two mirrors are not parallel to one another when the index is at zero on the arc. If the third adjustment is necessary the sextant possesses *index error*.

To test for index error, the index is set to zero. With the arc lying in the vertical plane the true and reflected images of a star by night or the horizon by day are then compared. If one image lies above the other, index error exists. By turning the tangent screw until the horizontal alignments of the images coincide, the amount of index error may be read from the arc. If the reading is positive the index error is described as being *on the arc*. If it is negative the index error is described as being *off the arc*. If index error is on the arc it will be necessary to subtract the error from all readings made with the uncorrected sextant. If it is off the arc index error will have to be added to all readings.

Index error is removed by making the third adjustment. This, like the first and second, is a screw adjustment. The screw for making the third adjustment is at the back of the cell which houses the horizon glass. The effect of turning this screw is to slew the horizon glass in its cell. The index-error adjustment screw is placed at the side of the vertical centre line of the mirror as viewed with the arc in the horizontal plane. In contrast, the adjustment screw for making the second adjustment is placed centrally on the edge of the housing farthest away from the plane of the arc.

Turning either screw at the back of the horizon glass cell usually affects both side error and index error. For this reason the second and third adjustments are made simultaneously. The usual method is to remove the existing side and index errors, and then to remove half the remaining side and index errors until a satisfactory adjustment is made for both.

If perpendicularity and/or side errors exist the three parts of the zig-zag ray due to the double reflection at the index and horizon mirrors will not be in the same plane. As a result of this all readings made with the uncorrected sextant will tend to be too high.

It is necessary that the part of the zig-zag ray which enters the observer's eye is parallel to the plane of the arc. If this is not so all readings will tend to be too high. This follows because the zig-zag ray will not lie in the same plane as that of the vertical circle through the observed heavenly body. Error due to this cause is called *collimation error*.

Collimation error may be due to the axis of the sextant telescope not being parallel to the plane of the arc, or it may be due to careless observing. When observing it is necessary that the observed image lies at the very centre of the field of view of the telescope. If this is not the case collimation error will result even if the telescope axis is set correctly.

The telescope is housed in a collar which, on some sextants, is adjustable. Two adjustment screws are fitted. The effect of slackening one of these screws and tightening the other is for the telescope axis to slew relative to the plane of the arc. This adjustment, when made to set the collar properly, is called the fourth adjustment.

In most modern sextants the housing of the telescope is permanently and properly fixed to the frame. Collimation error due to faulty housing is not, therefore, to be expected on such a sextant.

Before dealing with the manner in which collimation error may be detected, let us consider sextant telescopes.

The common practice at the present time is for manufacturers to supply a single telescope suitable for use with all observations. Such a telescope must inevitably be a compromise instrument. This follows because the requirements of a telescope for stellar observations are different from those for daytime or Sun observations.

For Sun observations the telescope should be capable of producing a large image of the Sun. The larger the image of the Sun (or Moon) the easier it is for the observer to make a sharp contact with the reflected image of the body's limb and the horizon.

The magnifying power of a telescope is related to the power of the eyepiece. The higher the power of the eyepiece the greater is the magnification. Theoretically there is no limit to the magnification of an image, but there are practical limits because the amount of light available to illuminate the magnified image is limited. The amount of light available for this purpose is related to the diameter of the object glass of the telescope, the amount being proportional to the area of the object glass. For any given object glass, the higher the magnification factor the smaller is the brightness of the magnified image.

For Sun observations there is usually an abundance of light, so that a small object glass is all that is necessary. The characteristic features of a sextant telescope designed for Sun observations are, therefore, high magnification and small object glass.

For observations made between the times of sunset and sunrise, light is not abundant. A large object glass is necessary therefore to capture as much light as possible in order that the horizon be rendered sharp and clear. The magnifying power of the eyepiece of the star telescope need only be small: stars never appear more than mere pinpoints of light even in the largest telescopes. The characteristic features of a star telescope for use with a sextant are, therefore, low magnification and large object glass.

The simplest type of telescope is the *astronomical telescope*. This type of telescope produces an inverted image and for this reason it is often called an *inverting* telescope. In contrast to the astronomical telescope, the *terrestrial telescope* is an *erecting* telescope, so-named because it is provided, in addition to the object glass and eyepiece, with an additional system of lenses designed to erect the inverted image produced by the object glass.

For convenience of use when observing stars, the sextant star telescope is an erecting telescope. On the other hand the Sun telescope is an inverting telescope. It is more difficult to use than a star telescope, and a large measure of skill is demanded of the observer when he uses an inverting telescope with his sextant. The time and effort spent in acquiring this skill is rewarded by the relatively high accuracy with which Sun altitudes may be measured by means of the inverting telescope compared with the relatively low accuracy of Sun altitudes when measured with a low-powered star telescope.

The importance attached to ensuring, when taking sights, that the image of the observed celestial body lies in the centre of the field of view of the telescope was stressed by John Hadley, the inventor of the reflecting quadrant, as far back as the mid-18th century. Hadley suggested the use of cross wires in the sextant telescope, by means of which the observer would be aided in respect of this important point. Hadley's suggestion seems never to have been noticed sufficiently.

The inverting telescope with which some sextant outfits are provided is usually fitted with two pairs of cross wires. When using such a telescope for measuring the Sun's altitude, the image of the observed limb should lie in the square formed by the intersecting cross wires, in order to eliminate the possibility of collimation error due to faulty observing.

The two pairs of cross wires serve to enable an observer to detect the possible presence of collimation error. To ascertain if the sextant possesses collimation error, the inverting telescope is shipped, and a pair of the intersecting cross wires is set parallel to the plane of the arc of the sextant. Two widely spaced stars are then chosen, and the reflected image of one is brought into coincidence with the direct image of the other, so that the two images lie on one of the pair of cross wires which are parallel to the plane of the arc. Having done this, the sextant is tilted slightly to bring one of the images on to the other cross wire of the pair. If the other image does likewise, and remains in coincidence, the sextant is free from collimation error due to faulty housing of the telescope. If however the two images separate, collimation error exists and it should be removed if possible by making the fourth adjustment.

When collimation error exists the measured altitude of a celestial body exceeds the required altitude by an amount which varies with the square of the angle of inclination of the line of sight to the plane of the sextant arc, and as the tangent of half the measured altitude. This may be proved as follows:

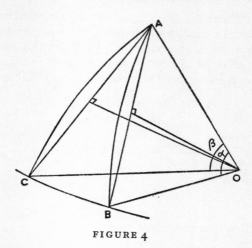

FIGURE 4

Fig. 4 serves to show that the arcs AB and AC of the sphere centred at 0 are proportional to the chords AB and AC, and that the ratio of the arcs or chords is proportional to that of the sines of half the angles subtended at the centre of the sphere, provided that the angle BAC is small.

Referring to Fig. 4:

$$\frac{\text{arc AB}}{\text{arc AC}} = \frac{\text{chord AB}}{\text{chord AC}}$$

$$= \frac{\frac{1}{2}\ \text{chord AB}}{\frac{1}{2}\ \text{chord AC}}$$

$$= \frac{\sin \alpha/2}{\sin \beta/2}$$

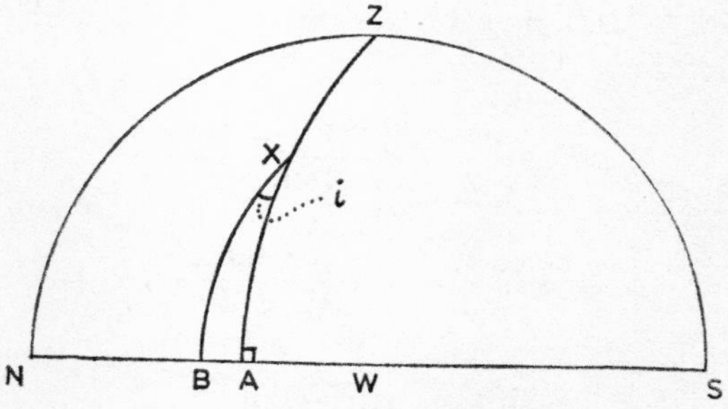

FIGURE 5

Figure 5 illustrates the celestial sphere projected on to the plane of the celestial meridian of an observer whose zenith is projected at Z. X is a star whose true altitude is arc AX indicated by θ. If the line of sight is inclined at an angle of i to the plane of the sextant arc, the measured altitude will be arc XB indicated by ϕ.

Now
$$\frac{\text{arc XA}}{\text{arc XB}} = \frac{\sin \theta/2}{\sin \phi/2}$$

but
$$\frac{\text{arc XA}}{\text{arc XB}} = \cos \text{AXB} = \cos i$$

therefore:
$$\cos i = \frac{\sin \theta/2}{\sin \phi/2}$$

Let collimation error be ϵ so that

$$\epsilon = \phi - \theta$$

Now
$$\cos i = \frac{\sin \theta/2}{\sin \phi/2}$$

and
$$1 - \cos i = 1 - \frac{\sin \theta/2}{\sin \phi/2}$$

i.e.
$$2 \sin^2 \frac{i}{2} = \frac{\sin \phi/2 - \sin \theta/2}{\sin \phi/2}$$

$$= \frac{2 \cos \frac{1}{2}(\phi/2 + \theta/2) \sin \frac{1}{2}(\phi/2 - \theta/2)}{\sin \phi/2}$$

Now $(\phi/2 + \theta/2)$ may be assumed to be ϕ and $(\phi/2 - \theta/2)$ is $\epsilon/2$, therefore:

$$2 \sin^2 \frac{i}{2} = \frac{2 \cos \phi/2 . \sin \epsilon/4}{\sin \phi/2}$$

i.e.
$$\sin^2 \frac{i}{2} = \frac{\cos \phi/2 . \sin \epsilon/4}{\sin \phi/2}$$

and
$$\sin \frac{\epsilon}{4} = \sin^2 \frac{i}{2} \tan \frac{\phi}{2}$$

Now ϵ and i are small angles and no significant error will result by assuming

$$\sin \frac{\epsilon}{4} = \frac{\epsilon}{4} \text{ radians}$$

$$\sin \frac{i}{2} = \frac{i}{2} \text{ radians}$$

so that:

$$\left(\frac{\epsilon}{4}\right)^{\circ} = \left(\frac{i}{2}\right)^{2} \tan \left(\frac{\phi}{2}\right)^{\circ}$$

$$\epsilon' = i^2 \tan \left(\frac{\phi}{2}\right)'$$

For values of $i = 1°$ and $\phi = 30°$, $60°$, $90°$, $\epsilon = 0.3$, 0.6 and 1.1 respectively.

It will be noticed that collimation error increases as the measured angle increases. It is for this reason that when using the method described above for detecting collimation error, the two stars chosen should be widely spaced, so that the effect of the possible inclination of the telescope is pronounced.

In addition to the four adjustable errors discussed above, a sextant may give false results because of one or a combination of several non-adjustable errors. The principal non-adjustable errors are called *centring error, prismatic error, shade error* and *graduation error*.

Centring error may result when the axis of the index bar does not coincide with the centre of curvature of the graduated arc. The magnitude of centring error varies with the position of the index bar on the arc. Special optical apparatus is required for detecting and measuring centring error. This apparatus is not normally available to a navigator. Centring error may be detected by comparing measured angles with their known values using a sextant which has been carefully adjusted for adjustable errors.

It is unlikely that centring error of any consequence will be found in a modern sextant. Careless use of the instrument, however, leading to wear on the bush at the pivot of the index bar, may lead to relatively large centring errors.

Prismatic error may result if the back and front surfaces of the index mirror or horizon glass are not parallel to one another.

The directions of the reflections from the index mirror and the silvered part of the horizon glass, and the direction of the direct ray that passes through the unsilvered part of the horizon glass, will be affected by refraction through any want of parallelism of the two surfaces of the mirrors.

Any prismatic effect of the horizon glass will affect the index error, and this effect may be eliminated by making the third adjustment. Prismatic effect of the index mirror, however, cannot normally be detected.

Shade error is the name given to error that may arise through the non-parallelism of the surfaces of the shades, or through lack of uniformity of the tint of the glass from which the shades are made. Shade error for any shade, or any given combination of shades, may be detected by comparing the values of fixed angles with and without the shades in place.

Graduation error may arise through faulty division of the arc, micrometer drum and/or Vernier scale. By using the Vernier principle graduation error, if it exists, may readily be detected. It would be unusual to find graduation error on a well-made sextant. In practice, therefore, this error is ignored.

SEXTANT ACCESSORIES

For the purpose of eliminating the effect of horizon glare, a *Nicol prism* is fitted to some sextants. This is simply a polarizing prism used like a telescope eyepiece. It is so fitted that when the telescope is screwed home in its collar, the polarizing plane of the prism is parallel to the plane of the arc of the sextant. Fitted in this way the polarizing plane is perpendicular to the horizontal when the sextant is used for measuring altitudes. The prism allows the 'extra-ordinary' ray to be transmitted to the eye, the intense glare being refracted upwards out of the prism.

Another useful device provided with some outfits is the *Wollaston prism* for use when taking star sights. The Wollaston prism, which is fitted between the index mirror and the horizon glass is, in fact, a pair of prisms in the form of two wedges of different thicknesses or different refractive indices, so that two distinct images of an observed star are formed.

When using the Wollaston prism the observer brings the two reflected images of the observed star to a position such that the horizon lies centrally between them. This device is of particular value for observing stars when the horizon is indistinct.

A device called a *lenticular* or an *elongating lens* is often fitted for use with star sights. This is a cylindrical lens fitted in a frame or cell and housed with the index shades. The purpose of the lenticular is to draw out the point image of the star into a line, thus facilitating the measuring of star altitudes.

An *artificial horizon* is used ashore when the sea horizon is not visible or when the observer's height of eye above sea level is not known. The traditional artificial horizon consists of a shallow trough filled with mercury to provide a bright horizontal reflecting surface. When using an artificial horizon of this type, the angle between the body and its reflection in the mercury is measured. Half this angle, after index error has been applied, is equal to the apparent altitude, regardless of the observer's height of eye above sea level.

The mercury horizon is useless for use on board ship unless the ship is perfectly steady. The slightest acceleration of the ship, especially through rolling or pitching, and the slightest vibration, would cause the mercury surface to tremble and become useless for observational purposes.

Many attempts have been made to provide the nautical astronomer with an artificial horizon that may be attached to the sextant. The most fruitful result of these efforts is the *Booth bubble horizon*, commonly found on sextants used by airmen. At the present time artificial horizons are seldom used by seamen.

CARE OF THE SEXTANT

The sextant is a precision instrument which demands careful handling in order to preserve its adjustments and prevent it from becoming damaged in any way. It should never be lifted other than firmly by the frame or handle. Should it be dropped or knocked sharply there is a big chance that it will be rendered useless for its purpose.

When not in use the sextant should be kept in its case and the case should be locked. It should be stowed securely in a suitable cupboard or on a chartroom shelf.

A great enemy of a sextant is moisture, especially salt water that may be allowed to remain on the mirrors or their cells. The sextant should be quite dry before stowing. A small quantity of silica gel, a substance which has a great affinity for water, should be kept in the sextant case. If this advice is followed, it may be necessary to dry out the silica gel occasionally in order for its effectiveness to be maintained. Care should be taken in drying or cleaning the sextant mirrors to prevent them from being scratched by grit that may be present in the cloth or chamois leather used for the purpose. The working parts of the sextant should be oiled lightly when necessary.

A sextant should not be exposed unnecessarily to the direct rays of the Sun. This treatment may lead to unsuspected error in measured altitudes.

After adjusting a sextant, the adjustments should hold good indefinitely. It should be borne in mind that frequent tinkering of the adjustment screws may result in the threads wearing, rendering the screws loose, with the resulting possibility of the adjustments being thrown out.

The following remarks on the sextant made by the renowned merchant seaman of the last century, Captain Lecky, are worth repeating.

'There is a proverb,' wrote Lecky, '"You should never lend to any one your horse, your gun, or your dog." It applies also to the sextant, only more so. Bear it in mind, dear boy.'

CHAPTER II

The Chronometer

A chronometer is an instrument designed for the purpose of keeping accurate time on board ship.

We have seen in Part I that the essential problem of finding longitude at sea is related to the comparison of local time with a standard time at the same instant. The difference between corresponding local and standard times is equivalent to the longitude of the local meridian eastwards or westwards of the standard meridian. The standard time used for this purpose is Greenwich Mean Time, longitude being measured eastward or westward from the Greenwich meridian.

The standard time with which computed local time is compared for the purpose of finding the ship's longitude is normally provided by a chronometer. A chronometer is an accurately made timekeeper the important features of which are:

1. The energy derived from the wound mainspring is transmitted through a train of gears and communicated by the escapement to the balance, in a uniform manner by means of a variable lever device known as the *fusee.*
2. Compensation for temperature changes is achieved by means of a *bi-metallic balance wheel.*

The amount by which a chronometer gains or loses during 24 hours of Mean Time is called the chronometer's *daily rate.* The rate of a chronometer is related to temperature. A rise in temperature causes the rate of a timekeeper fitted with a simple *uncompensated* balance wheel to retard, whereas a fall in temperature results in an accelerated rate. The purpose of the *compensated* balance is to correct this defect.

The effects of a change in temperature are a change in the

tension of the balance spring and a change in the moment of inertia of the balance wheel due to a change in the distribution of the mass of the wheel.

The tension in the balance spring varies directly as temperature, whereas the moment of inertia of the balance wheel varies as the square of the temperature. Accordingly, there are two, and only two, temperatures at which temperature compensation is correct. Chronometers are constructed so that they are correctly compensated for temperature at two standard temperatures, viz. 45°F and 75°F (6°C and 24°C). At temperatures between the standards a compensated chronometer should gain: at other temperatures it should lose.

The chronometer is mounted on gymbals so that it maintains a horizontal position despite rolling and/or pitching of the ship. It should be housed in a permanent position in the chronometer box in a locker which should be dustproof and insulated to offset the effects of rapid changes in temperature.

Before the advent of radio time signals the chronometer provided the only satisfactory means of finding G.M.T., at any time of the day or night. The rigorous routine related to the management of a chronometer on board ship, which was formulated in byegone days, has persisted to the present time. The rules associated with the checking, winding and handling of the precious timekeeper are still practised with an observance almost ritual.

Nowadays, when radio time signals are available at any time of the day, the chronometer has lost some of its former glory. Moreover, improvements made in watchmaking have resulted in the availability of portable watches having a sufficiently high degree of accuracy to obviate the need for the more expensive chronometer.

The standard chronometer, which will run for about 56 hours after winding, is described as a *two-day chronometer*. To ensure a regular routine for winding, so reducing the possibility of an uneven rate, a standard chronometer should be wound daily. On merchant ships the Second Mate is usually entrusted with the care of the ship's chronometer, and it is part of this officer's duty to wind the chronometer at the same time every day. When the ship is at sea, the regular routine of the ship makes it unlikely for this important task to be overlooked. When the ship is in port, however, distractions due to a variety of causes often result in the chronometer remaining unwound. Allowing a chronometer to stop through carelessness is an offence which every self-respecting Second Mate lives in fear of committing.

In order to wind the chronometer, the instrument is first turned on its side, and the guard covering the keyhole is slid back. The key is then inserted and the instrument wound until the key butts. This requires about seven half turns. The winding key, called a tipsy key, is designed so that if it is turned in the wrong direction the winding mechanism will not be stressed. A small dial on the chronometer face serves to indicate whether the chronometer is fully or partly wound.

It is very important that the chronometer is not allowed to stop, especially on small vessels not fitted with radio. The escapement of a chronometer, unlike that of an ordinary watch, is not self-starting. Should the chronometer stop, therefore, it will be necessary, after winding it, to turn it sharply to right or left in order to set it going. Should it be necessary to re-set the hands, this should be done with the tipsy key before winding the chronometer. On no account should the hands be turned other than by means of the key, which is designed to fit over the spindle to which the hands are fitted.

A chronometer should be cleaned and oiled regularly at intervals of two or three years. This work should be undertaken by or through a nautical instrument supplier.

Before transporting a chronometer from or to a ship it will be necessary to wedge the balance wheel using two thin cork wedges for the purpose. In order to do this the instrument is removed from its box and the glass face unscrewed. The brass case is then grasped at the bottom and, with the fingers of the left hand placed around the top edge, the case is inverted and lifted off the working part. The key must be used to ease the working mechanism from the case should it be necessary to force it. After inserting the wedges the chronometer should be reassembled in its box and clamped in its gymbals. The box containing the instrument should be carried carefully and it should not be allowed to suffer jolts or shocks.

When the ship is to be laid up for a long period the chronometer should be placed under the care of a nautical instrument supplier.

When a chronometer is delivered to a ship it is customary for the nautical instrument supplier to furnish the navigator with a *Chronometer Rate Certificate*, on which is written the error and rate of the instrument. During the time when the ship is in service it is important that a careful check be made, and a record kept, of the chronometer rate. If this becomes erratic or unduly large, chronometer times used for astronomical purposes should be treated with caution. A British Admiralty chronometer is regarded as being unfit for navigational purposes when its daily rate is erratic or when its daily rate exceeds six seconds per day.

A discussion on time signals and their use for checking chronometers is given in Part IV, Chapter II.

CHAPTER III

The Nautical Almanac

The word *almanac* comes from the Arabic *al manakh* meaning the calendar. An almanac contains astronomical data, and a *nautical* almanac provides the seaman with the astronomical data he needs for position-finding using the methods of nautical astronomy. It is a navigational instrument of the first importance for without the information it provides the navigator's sextant and chronometer would be of little use.

The earliest almanacs used for astronomical navigation by the Portuguese and Spanish navigators during the early period of the Golden Age of Discovery, contained little more than that necessary for finding latitude. It was not until the 18th century that satisfactory methods for finding longitude at sea were available to the mariner, and it was during this time that the first British *Nautical Almanac* was published. This almanac, which was first published in 1766 for the year 1767, contained astronomical data useful not only for finding latitude at sea, but also for finding longitude.

The principal method for finding longitude at sea during the century following the appearance of the first British *Nautical Almanac* was the 'lunar method'. This method involved measuring the angle between the Moon and a selected star, planet or the Sun, and, after clearing the measured angle for the effects of refraction and parallax, comparing it with predicted *lunar distances* given in the almanac against G.M.T. Now that G.M.T. is readily available to a navigator by means of his chronometer or a radio time signal, it is no longer necessary to find longitude at sea by means of the relatively complex lunar distance method, and predictions of lunar distances are no longer available.

The essential problem in modern nautical astronomy is concerned with the relationship between the position of a celestial body at any given instant of time using the horizon and equinoctial systems of co-ordinates respectively. The data contained in the *Nautical Almanac* facilitates this problem.

The British *Nautical Almanac* is identical in contents with the United States' *Nautical Almanac*. It is published jointly by H.M. Nautical Almanac Office and the United States' Nautical Almanac Office. The British and United States' Governments make available to any nation the data contained in the almanac in a form suitable for direct photographic reproduction. At the present time the Brazilian, Danish and Norwegian almanacs are produced under this arrangement.

The principal part of the British *Nautical Almanac* consists of astronomical data by means of which the declination and Greenwich hour angle (G.H.A.) of every celestial body used in nautical astronomy may be ascertained for any instant of G.M.T. The declination of an observed body is required to form the side PX (polar distance of observed body) of the astronomical triangle. The G.H.A's of the navigational bodies are given so that a local hour angle (L.H.A.) or a longitude may be found from the relationship:

$$\text{L.H.A.} * \sim \text{G.H.A.} * = \text{longitude}$$

where * represents any celestial body.

The principal part of the almanac, therefore, consists of tables giving predicted positions of the Sun, Moon, navigational planets, and the First Point of Aries, using the equinoctial system of co-ordinates (Declination and G.H.A.), against G.M.T. Such an astronomical table is called an *ephemeris.*

Tabulated values of G.H.A. and declination, given to an accuracy of 0·1', are given for every integral hour of G.M.T. for the whole year, and special interpolation tables are provided so that values for any G.M.T. may be found.

A table giving sidereal hour angle (S.H.A.) of each of the navigational stars is provided. By applying the S.H.A. of a star to the G.H.A. ♈ for any given instant, the G.H.A. of the star for the same instant may be found.

In addition to the ephemerides, the *Nautical Almanac* contains a great deal of other information of use to the nautical astronomer. Included in this information are: a calendar of religious and civil holidays, eclipse information, notes on planets, times of Sunset and Sunrise, times of Moonset and Moonrise, times of twilight, equation of time, times of meridian passage of Sun and Moon, table of standard times, Pole Star tables, star charts and altitude correction tables. In addition to this information several pages of the *Nautical Almanac* are given over to an explanation of the principles, arrangement and use, of the almanac. Every nautical astronomer using this almanac should be thoroughly familiar with this explanation.

Each pair of facing pages, known as *daily pages*, contains the ephemeral data, arranged vertically, for three days. On the left-hand daily page successive columns give G.H.A. ♈ and G.H.A. and declination of each of the four navigational planets Venus, Mars, Jupiter and Saturn, followed by a table giving the name, S.H.A. and declination, of each of fifty-seven selected stars. At the bottom right-hand corner of this page the S.H.A's and G.M.T's of meridian passage of the navigational planets over the meridian of Greenwich are given.

The right-hand daily page contains the Sun and Moon ephemerides. Twilight information, times of Sunset, Sunrise, Moonset and Moonrise, are also given on this page. At the bottom right-hand corner of the right-hand daily page will be found the equation of time given for 00 hr and 12 hr for each of the three days, the G.M.T. of the Sun's meridian passage at Greenwich, and the G.M.T. of the Moon's upper and lower transits at Greenwich, together with the age and phase of the Moon.

The special interpolation tables provided in the *Nautical Almanac*, which are placed near the end of the book, are printed on tinted paper so that they are readily located.

The altitude correction tables are given at the front and back of the book. Those for the Sun, stars and planets are given on the inside of the front cover and the facing page; and those for the Moon are given on the inside of the back cover and the facing page. A dip table is duplicated in both sets of correction tables.

The *Nautical Almanac* dip table is a critical table. Values of dip are calculated from the formula:

$$\text{dip in minutes} = 0.97\sqrt{H}$$

where *H* is the height of the observer's eye in feet above sea level. The dip table is arranged so that at a critical entry the upper of the two values is to be taken.

The Sun's altitude correction table, given on pages A2 and A3 of the almanac, is the combined correction table for mean refraction, mean semi-diameter for each of two periods (October to March and April to September), and mean solar parallax. The table is arranged critically in two parts, one for altitudes greater than about 10°, and the other for altitudes between 0° and 10°. Corrections are given for both lower and upper limb observations, those for the latter including an irradiation correction of −1·2′.

The argument with which the Sun's altitude correction table should be entered is apparent altitude. This is the arc of a vertical circle contained between a celestial object and the sensible horizon. To convert the observed altitude, after index error (if any) has been applied, into apparent altitude, the dip correction must be applied to the former.

The mean refraction correction given in the altitude correction tables is based on a theory of refraction by B. Garfinkel (see the *Astronomical Journal*, Volume 50, 1944), and is given for a *standard atmosphere* having a temperature of 10°C and a sea level pressure of 1010 millibars. For *non-standard* conditions it is necessary to apply an additional correction which may be found from page A4. This correction is particularly important when the altitude of the observed body is small.

The altitude corrections for planets are the same as those for stars; but for Venus and Mars additional corrections for parallax and phase are necessary. *Phase correction* is necessary because the direction of the actual centre of a planet is different from that of its apparent centre. Phase correction is insignificant except for the relatively near planets Venus and Mars. The phase and parallax corrections depend upon the relative positions of the Sun and the planet, and upon the relationship between the planet and the horizon. The corrections given in the altitude correction table apply to twilight observations. Should Venus be observed when the Sun is above the horizon, the phase and parallax correction should be computed directly, using the altitude, phase angle (the angle between the vertical circle through the planet and the direction of the Sun from the planet), and constants p and k which are given in the *Nautical Almanac*.

The altitude correction tables for Sun, stars and planets, given on page A2, are critical tables: those on page A3 are non-critical.

The Moon altitude correction table is a combined table giving corrections for mean refraction, semi-diameter, augmentation and parallax. The correction is in two parts, the argument for the first correction, which is given in the top part of the table, is apparent altitude. Those for the second part, which is given in the lower part of the table, are apparent altitude, limb, and horizontal parallax. The table is designed so that all corrections are positive, but 30′ are to be subtracted from altitudes of the Moon's upper limb.

The practical use of the *Nautical Almanac* is described in Part IV, Chapter I.

CHAPTER IV

Inspection and Short-Method Tables

The term *inspection table* applies to a table giving direct solutions of the PZX triangle. The earliest navigation inspection tables were prepared by the great astronomer Cassini in about 1770. Cassini's *horary tables* were designed to give local hour angle as respondent against arguments latitude, declination and altitude. Other early horary tables are those of Lalande published in 1793, and Thomas Lynn published in 1827.

Lynn's tables, like those of Cassini's and Lalande's, were designed for Sun observations. They give tabulated solutions for hour angle for each integral degree of latitude from 0° to 60°, of declination from 0° to 24° N. and S., and of altitude from 0° to 60°.

The early horary tables never became popular, largely on account of the tedious interpolation that was necessary in order to obtain accurate results from their use. Seamen have never taken kindly to interpolation, and when triple interpolation is needed, as is the case with the early horary tables, most seamen prefer to solve their PZX triangles using the direct methods of spherical trigonometry.

Improvements in horary tables made towards the end of the last century were related to improved methods in interpolation. Notable in this respect are the *Chronometer Tables* designed by Percy L. H. Davis and published in 1897.

Following the introduction of the intercept method of sight reduction, which came into general use in the Royal Navy in the early part of the present century, attention was directed to the task of designing inspection tables to give solutions of zenith distance (or altitude), against arguments latitude, declination and hour angle. The first inspection tables designed for use with the intercept method were those published in 1907 under the authorship of the Royal Naval Instructor Frederick Ball.

Ball perceived that since in the intercept method the observer may use any position near the D.R. position of his ship, it is possible to choose the position for solving a sight so as to have an integral number of degrees of latitude and hour angle. Interpolation, using this simple principle, was necessary, therefore, only for odd minutes of declination: the correction for declination when using Ball's tables was effected by means of a small supplementary table.

The *Altitude–Azimuth Tables* of Percy L. H. Davis were first published in 1917. These tables give both altitude and azimuth against latitude and declination (each to an integral degree), and local hour angle at intervals of eight minutes of time. Davis's tables became very popular, especially for finding azimuths, for star identification, and for planning sights. The relative difficulty of interpolation in these, as well as in most of the early inspection tables, spelt their doom, and they are seldom used at the present time.

The growing need for a quick and simple method of sight reduction has resulted in the publication of a comprehensive set of inspection tables. These, the title of which is *Tables of Computed Altitude and Azimuth*, were published by the United States' Government in 1946 in nine volumes. They are generally known as H.O. 214 and are similar in design to Davis's *Chronometer Tables* referred to above.

The United States' Hydrographer permitted the reproduction of H.O. 214 by the Hydrographic Department of the British Admiralty. The corresponding British tables, known as H.D. 486, were published in six volumes, each volume embracing 15° of latitude, in 1953.

The *Tables of Computed Altitude and Azimuth* formed at the time the most comprehensive set of tables of altitude and azimuth in existence. An important feature of the tables is that there are no precepts connected with their use. The entering arguments are latitude, declination and hour angle, and the respondents are altitude to $0'\cdot 1$ and azimuth to $0°\cdot 1$.

For each degree of latitude there are 24 pages of tables. Declination entries are given at half-degree intervals, each entry being given at the top of each of eight columns per page. The declination entries cater for all celestial bodies of navigational importance having declinations of up to 74° 30′. The extreme left-hand column on each page is labelled H.A. meaning local hour angle east or west of the local meridian. H.A. entries are given for every integral degree from 0° to the top to the maximum value at which the altitude of the observed body is 5° or more.

At an opening the left-hand page gives tabulated altitudes and azimuths for cases in which the latitude and declination have the same name. The right-hand page applies to cases in which the latitude and declination have contrary names.

In each declination column there are four vertical columns of figures. The figures in the first of these are in bold-faced type and are altitudes. Those in the fourth are azimuths. The second and third columns are labelled Δd and Δt respectively. Δd is the change in altitude for a change of 1′ in the declination. Δt is the change in altitude for a change of 1′ in the hour angle.

The standard (and simplest) method of using H.D. 486 is to work the sight using a chosen position the latitude of which is an integral number of degrees, and the longitude of which is such that the hour angle of the observed body is also an integral number of degrees. By so doing the required altitude and azimuth may be found with interpolation for declination only. The standard method of sight reduction involves the following steps:

1. Find G.H.A. and declination of observed body from the *Nautical Almanac*.
2. Choose a latitude nearest in integral degrees to the ship's estimated latitude.
3. Choose a longitude nearest to the ship's estimated longitude such that when it is applied to the body's G.H.A. a L.H.A. of an integral number of degrees results.
4. Enter tables with latitude, L.H.A. and nearest tabulated declination to that given in *Almanac*.
5. Extract altitude and azimuth and Δd, noting whether Δd is increasing or decreasing by inspection of neighbouring declination columns.

6. Multiply Δd by the difference in minutes of arc between the declination from the *Almanac* and that used to enter the tables. (A multiplication table is provided to facilitate this process.)

7. Apply the Δd correction to the tabulated altitude.

8. Name the azimuth and find the required intercept.

The method is exemplified as follows:

EXAMPLE: Using the extract from H.D. 486, find the calculated altitude and the azimuth of a celestial body whose declination is 12° 24′ N. The chosen latitude is 42° 00′ N. and the chosen longitude yields an H.A. of 6° W.

Lat 42°   Declination same name as latitude

H.A.	12° 00′				12° 30′			
	Alt.	Δd	Δt	Az	Alt.	Δd	Δt	Az
	° ′			°	° ′			°
4	59 47·9	99	11	172·2				
05	59 41·1	99	16	170·3	60 10·8	99	14	170·1
6	59 32·8	99	16	168·4	60 02·4	99	16	168·2
7	59 23·1	98	19	166·5	59 52·6	98	19	

Enter table with latitude = 42° N.

H.A. = 6°

declination = 12° 30′ N.

(N.B. Lat and dec same name)

Altitude = 60° 02·4′ Δd = −99

Δd corr =   −5·9

Altitude = 59° 56·5′   Az = N. 168·2° W.

The principal disadvantage of using the standard method is that each sight of a set requires the use of a chosen longitude different from that of each of the remainder. This leads to complication in plotting, for each intercept will have to be plotted from a different position.

H.D. 486 is readily adapted for use with an estimated position; but when so used, double or triple interpolation will be necessary. Instructions for use are given in the introduction to the tables.

An interesting and valuable set of sight reduction tables, first published in 1953, are those known as A.P. 3270. The Nautical Almanac Office of the U.S. Naval Observatory and H.M. Nautical Almanac Office co-operated in the design and preparation of these tables. They were originally planned in the United States where they are published as H.O. 249. These tables, which were designed essentially for air navigation, are in three volumes. Volume I contains precomputed altitudes to the nearest minute, and azimuths to the nearest degree, of six selected stars for each integral degree of latitude and L.H.A. ♈. Volumes II and III, which are identical in character—Volume II covering latitudes 0°–39°, and Volume III covering latitudes 40°–89°—are designed for solving Sun-, Moon- and Planet-sights. They may also be used for star-sights so long as the declination is between 0° and 29° N. or S., this being the range of declination covered by each of the two volumes.

Volume I of A.P. 3270 serves admirably not only for reducing star-sights but also for planning twilight observations. All the data relating to six well-placed stars for each degree of latitude are presented on two facing pages of the tables. The page heading is latitude and the altitudes and azimuths of the six stars are given in vertical columns against L.H.A. ♈ as vertical argument. This arrangement facilitates seeing at a glance for any given L.H.A. ♈ the approximate altitudes and azimuths of the six selected stars. The approximate altitudes can thus be set on the sextant and each star observed in turn on its known bearing.

What may be regarded as the *tour de force* in connection with inspection tables are the newly published (1967) *Sight Reduction Tables for Marine Navigation.*

Although H.O. 214 (H.D. 486) is not without imperfections it was not, until recently, thought that the enormous task of recomputing, proof-reading and rearranging the tabular material to improve the efficiency of the tables warranted the expense and labour that would be entailed in replacing it. However, with the availability of electronic computers and other improvements which make for high-speed setting and checking, the position has changed. Accordingly the U.S. Naval Oceanographic Office (formerly the U.S. Hydrographic Office) decided to undertake a design study for a possible replacement of H.O. 214 (H.D. 486). The result has been the publication of *Sight Reduction Tables for Marine Navigation*, the production of which has been the co-operative effort of the U.S. Naval Oceanographic Office, the Nautical Almanac Office of the U.S. Naval Observatory and H.M. Nautical Almanac Office.

In producing this new work the aim was to provide the mariner with tables by which, with conventional methods of observation and altitude correction, the highest precision possible is attainable.

The tables appear in six volumes. Altitudes to the nearest $0'\cdot1$, and azimuths to the nearest $0°\cdot1$ are tabulated for all combinations of latitude, L.H.A., and declination at uniform intervals of $1°$. Interpolation is required only in respect of declination—specially designed tables being provided for this purpose.

Inspection tables provide what are sometimes called *tabular methods* of sight reduction. Two other categories of methods of sight reduction are *direct methods* and *mixed methods*.

In the so-called direct methods of sight reduction nothing more than a number of single entry tables are required, and interpolation is completely avoided. In the tabular methods considerable interpolation is necessary unless the table is abnormally bulky and therefore expensive. The H.D. 486 inspection tables discussed above, for example, form a veritable library of six volumes, each volume containing about 360 large pages. The so-called mixed methods include those in which the PZX triangle is split by dropping a perpendicular great circle from one of its corners—usually X or Z—on to the opposite side: and solving the PZX triangle for altitude or hour angle and/or azimuth. The tables designed for mixed methods are usually called *short-method tables*. We shall discuss some of the mixed methods and their associated short-method tables in this chapter, reserving a discussion on some of the direct methods for Part IV, Chapter IV.

Numerous attempts have been made and a great deal of energy expended by many brilliant mathematicians in producing short methods of nautical astronomy by which the seaman may be aided in solving his PZX triangles. Notable amongst these attempts was that made by Sir William Thomson (later Lord Kelvin), who published what may be regarded as the first short-method table for astronomical navigation.

Sir William Thomson is credited with being the first to apply the well-known device of dividing an oblique spherical triangle into two right-angled spherical triangles, to facilitate solution, to the needs of nautical astronomy in respect of finding longitude at sea. Thomson's remarks on solving PZX triangles are interesting:

'. . . When we consider the thousands of triangles calculated daily among all the ships at sea we might be led for a moment to imagine that every new calculation is merely a repetition of one already made. But this would be a prodigious error: for nothing short of accuracy to the nearest minute in the use of the data would thoroughly suffice for practical purposes. Now there are 5400' in 90° and therefore there are 5400^3 or 157,464,000,000 triangles to be solved each for a single angle. Even with an artifice such as that to be described, for utilizing solutions of triangles with their sides integral numbers of degrees, the number to be solved (being 90^2) or 729,000, would be too great, and the tabulations of the solutions would be too complicated (on account of the trouble of entering for the three sides) to be convenient for practice; and tables of this kind which have already been actually calculated and published (as for example Lynn's Horary Tables of 1827) have not come into general use.' (*Nautical Magazine*, 1871).

It occurred to Thomson that by dividing the problem into the solution of two right-angled triangles, the ship's position could be found without recourse to calculation. Thomson's method, in which the PZX triangle is split into two right-angled triangles by dropping a perpendicular from the observed body on to the observer's celestial meridian, is the first of numerous methods in which the same technique is employed.

By clever design Thomson's tables, which were published in 1876, are amongst the briefest of those produced for the same purpose. Despite the ingenuity and mathematical talent of their inventor, and largely because of the complexity of the rules for using them, Thomson's tables were not a success. It is the complexity of the rules for using them that is the principal drawback of most short-method tables. However, once the rules are mastered, and sufficient practice has been obtained in the use of any of most of the short-method tables, a considerable saving of time, as compared with that needed for most of the direct methods, and even some of the tabular methods, is effected.

Short-method and inspection tables are not generally used in the Merchant Navy, the officers of which seem to cling to the cumbersome direct methods using logarithms. The Board of Trade Examiners of Masters and Mates appear not to recognize any but direct methods for sight reduction, and this, seemingly, is a factor which detracts many Merchant seamen from learning short methods.

The late Vice-Admiral Radler de Aquino ranks as one of the greatest navigators, and a foremost authority on short-method tables, of his times. He first published his famous *Altitude and Azimuth Tables*, which he described as being 'the simplest and readiest in solution', in 1907. The remarks with which he prefaced the English editions of his early tables are interesting:

'Attention!'

'Would you ever think of going to the trouble of calculating the elements of the *Nautical Almanac* . . . when the Nautical Almanac office tabulates these data for you? Would you ever think of working out your D.R. by means of formulae and logarithms when the *Plane Traverse Table* facilitates the direct solution of all problems related thereto?

Why then go to the trouble to solve the astronomical triangle by means of complicated formulae and logarithms when we have tabulated its elements in our *Altitude and Azimuth Tables* (spherical traverse tables) and have given the simplest and readiest methods for solving all problems related thereto?'

Aquino's tables were designed for use with the intercept method. Aquino's aim was to provide a means of dispensing with logarithms and with only a limited amount of interpolation to determine the zenith distance and azimuth of an observed body for an assumed position.

Aquino's *Altitude and Azimuth Tables* are based on the splitting of the PZX triangle into two right-angled triangles by dropping a perpendicular great circle from the observed body on to the observer's celestial meridian. They form, in effect, a spherical traverse table for two right-angled triangles PMX and ZMX, M being the foot of the perpendicular from X on to the observer's celestial meridian. The side MX, being common to both triangles, acts as a link.

An interesting navigational method based on the Marcq St. Hilaire principle was published in 1920 by the Hydrographic Department of the Japanese Navy. This publication, entitled *New Altitude and Azimuth Tables between 65° N. and 65° S. for the Determination of the Position Line at Sea*, was the work of S. Ogura of the Japanese Navy.

In using Ball's or Aquino's *Altitude and Azimuth Tables*, the navigator makes use of an integral number of degrees in his assumed latitude and the local hour angle. The same technique is used in Ogura's method.

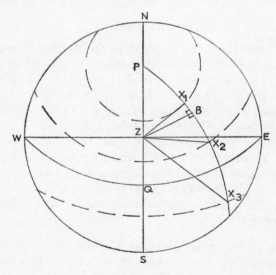

FIGURE I

In Ogura's method the PZX triangle is split by a perpendicular great circle from Z on to the opposite side or side produced, as in Fig. 1.

The declination of the point B at the foot of the perpendicular from Z is denoted by K.

When the latitude of the observer and the declination of the observed body have contrary names the arc BX is equal to the sum of K and the body's declination. When the latitude and declination have the same name the side BX is equal to the difference between K and the declination of the body. In general:

$$BX = (K \pm d)$$

where d is the declination of the observed body.

Referring to Fig. 1:

In triangle PZB (known as the *Time triangle*) we have, from Napier's rules:

$$\tan PB = \cos P \tan PZ$$

Inverting this we have:

$$\cot PB = \sec P \cot PZ$$

i.e.
$$\tan K = \sec P \tan \text{lat} \qquad (1)$$

Also: $\sin BZ = \sin P \cos \text{lat}$
from which:

$$\text{cosec } BZ = \text{cosec } P \sec \text{lat} \qquad (2)$$

In triangle BZX (known as the *Altitude* or *Zenith Distance triangle*) we have, from Napier's rules:

$$\cos ZX = \cos BZ \cos BX$$

from which:

$$\sec ZX = \sec BZ \sec (K \pm d)$$

or:
$$\log \sec ZX = \log \sec BZ + \log \sec (K \pm d)$$

Ogura's method marked a distinct advance in nautical astronomy. All that was necessary to find the calculated zenith distance when using the intercept method, was to lift A and K

from a specially designed table, combine K with the declination of the observed body, then add the log secant of $(K \pm d)$ to A (which latter element is log sec BZ) to give the log secant of the required zenith distance.

The principal feature of Ogura's method is its conciseness, and his table involves complete freedom from interpolation.

H. B. Goodwin, a well-known writer on navigation and nautical astronomy during the early part of the present century, suggested that Ogura's table might well be included in the common nautical table collections. Goodwin's suggestion seems to have borne fruit, for in 1924 a table entitled *Short Method for Zenith Distance* was published in Norie's *Nautical Tables* as the *A and K Tables*.

The *Altitude-Azimuth Table* contained in the present Norie's collection is designed to give the azimuth of the observed body as well as its calculated altitude. The reader is referred to the explanation given in Norie's tables.

The well-known *Hughes' Tables for Sea and Air Navigation*, first published in 1938, are designed on Ogura's method. These tables, the tabulated quantities of which were mechanically computed, was skilfully designed by the late Doctor L. J. Comrie of the British Nautical Almanac Office.

An interesting short-method table designed specifically for the air navigator, but now adapted for marine use, is the *Rapid Altitude and Azimuth Tables* by Myerscough and Hamilton. These tables are similar to Ogura's, the PZX triangle being divided by dropping a perpendicular great circle from the zenith on to the side PX or PX produced, and are designed for use with a chosen position the latitude of which has an integral number of degrees, and the longitude of which yields a L.H.A. having an integral number of degrees.

The data in the Myerscough and Hamilton tables are laid out in a clear manner in order to facilitate the solution of the altitude and azimuth with the minimum of effort. The azimuth table is based on the *A, B and C Table* principle which is described in the following chapter.

Most short-method tables make use of a position having an integral number of degrees in its latitude and a longitude which produces an integral number of degrees in the hour angle. A popular table of the present time is that compiled by J. C. Lieuwen by order of the Netherlands Ministry of Marine. Lieuwen's tables are based on Ogura's principle and are designed for use with a D.R. position.

The Netherlands' edition of Lieuwen's tables, after the tables had been tested by a committee under the chairmanship of the Chief of the Netherlands Hydrographic Office, has been made compulsory for all schools and colleges of the Netherlands Royal Navy and Mercantile Marine. An English edition has been published by Messrs. George Philip and Son Limited.

CHAPTER V

Miscellaneous Nautical Tables
and Instruments

In this chapter we shall discuss some of the tables and instruments used by nautical astronomers, other than the essential tables and instruments described in other chapters of this book. In particular we shall discuss azimuth tables and diagrams, star globes and star finders, and slide rules designed for nautical astronomy.

For the purposes of checking compasses a navigator employs the true azimuth of a celestial body. The true azimuth of a heavenly body is a measure of the arc of the horizon contained between the observer's celestial meridian and the vertical circle through the body. This angle, when compared with the compass bearing at any given time, enables the navigator to ascertain the error of his compass for the heading of his ship at the time.

As well as providing the means for checking compasses, the true azimuth of a celestial body whose altitude is observed for the purpose of position-line navigation serves to enable the observer to ascertain the direction of a position line—this being at right angles to the azimuth of the observed body at the time of the observation.

In pre-position line days the purpose of finding the true azimuth of a celestial body (particularly that of the Sun) was geared almost entirely to the need for checking compasses. With the advent of iron and steel ships the need for comprehensive and simple azimuth tables became acute, and the earliest azimuth tables were designed essentially for use in connection with magnetic compass checking.

Captain Thomas Lynn, famous for his inspection tables which

are mentioned in Chapter 4, produced a large azimuth table as early as 1829. Lynn's azimuth table gave azimuths of the Sun against latitude of the observer and declination and altitude of the Sun.

The most famous of all azimuth tables is the monumental work of Staff Commander John Burdwood, R.N., *Tables of Sun's True Bearing or Azimuth*, devised by Burdwood and first published in 1852. These were enlarged, first by Burdwood himself, and later by Captain John E. Davis, R.N., and his son Percy L. H. Davis of the British Nautical Almanac Office. Burdwood's and Davis's azimuth tables are designed to give azimuth against latitude of the observer and declination and hour angle of the Sun. They cover all latitudes between 64° N. and 64° S.

With the advent of position-line navigation, which dates from the middle of the 19th century, existing azimuth tables afforded a ready means for finding the azimuth, and thence the direction of a position line obtained from an altitude observation of a celestial body.

The popularity of Burdwood's and Davis's tables are matched only by the celebrated *A, B and C Tables*. These tables are used extensively by navigators of all nationalities, and in particular by those who use the direct methods of sight reduction in preference to the less troublesome and less time-consuming mixed or tabular methods.

The *A, B and C Tables* found in collections of nautical tables, such as those of Burton's and Norie's, are of great practical utility. The history of these interesting tables dates from 1845 when Lieutenant (later Admiral) L. G. Heath, R.N., invented the original *A and B Tables* which are described in the *Nautical Magazine* of 1846.

The original purpose of the *A and B Tables* was to facilitate finding the noon longitude, as soon as the noon latitude had been obtained from a meridian altitude observation of the Sun, from the a.m. Sun sight.

Heath's *A and B Tables* were based on the relationship:

Error in longitude due to error in latitude
$$\propto \tan \text{lat} \cot \text{H.A.} \pm \tan \text{dec} \csc \text{H.A.}$$

Values of tan lat cot H.A. were tabulated as A against arguments latitude and hour angle; and values of tan dec cosec H.A. were tabulated as B against arguments declination and hour angle.

By combining A and B, the error in longitude due to an error of one minute error in latitude is found. This error in longitude is commonly called the *longitude correction* or *longitude factor*.

$$\text{Longitude correction} = A \pm B$$

A short while after the first appearance of Heath's *A and B Tables*, a Royal Naval Instructor named J. N. Laverty published, for private circulation, a small work in which he included a table from which the longitude factor could be lifted using arguments latitude of observer and azimuth of observed body.

Laverty's longitude correction table is based on the relationship:

$$\text{Longitude correction} = \cot \text{Az} \sec \text{Lat}$$

Heath's *A and B Tables*, enlarged by Blackburne and Lecky, and Laverty's longitude correction table enlarged by Johnson and Lecky, were first combined by a well-known nautical teacher of his day and an editor of Norie's nautical tables and epitome, named W. H. Rosser. Rosser's *A, B and C Tables* first appeared in the Norie's collection in 1889.

A, B and C factors may be derived from the four-parts formula of spherical trigonometry. This formula applied to the astronomical triangle reduces to:

$$\cot Z \sec \phi = -\tan \phi \cot h + \tan d \operatorname{cosec} h$$

where Z = azimuth of observed body

ϕ = observer's latitude

d = declination of observed body

h = H.A. of observed body

Now: $\cot Z \sec \phi = C$

$$-\tan \phi \cot h = A$$

$$\tan d \operatorname{cosec} h = B$$

therefore:

$$A + B = C$$

By combining A and B factors and entering with this combination as one argument C in the *A, B and C Tables*, the azimuth of the observed body may be lifted using latitude as the second argument.

The azimuth of a celestial body is the angle at the observer's zenith contained between the observer's celestial meridian and the vertical circle through the body. It is an angle in the PZX triangle and its value lies between 0° and 180°. If the latitude of the observer and the declination of the body have different names, the azimuth of the body is always greater than 90°. The azimuth of a body is always less than 90° when the declination of the body is of the same name as but of greater magnitude than the observer's latitude. When latitude and declination have the same name, but the declination is smaller than the latitude, the azimuth of a celestial body may be less or greater than 90°. Azimuth is always named according to latitude and hour angle.

The *rising amplitude* of a celestial body is a measure of the arc of the horizon between the east point of the horizon and the body when it rises out of the horizon. The angle between the west point of the horizon, and an object when setting is called the object's *setting amplitude*. The amplitude of a body is always named from East or West and takes the name of the body's declination.

An *amplitude table* is commonly found in collections of nautical tables. In amplitude tables the amplitude of a body is tabulated against latitude of observer and body's declination, assuming the body to have a true altitude of 00° 00'. The true altitude of a celestial body is 00° 00' when it lies on the celestial horizon.

Tabulated amplitudes in amplitude tables are computed from the relationship:

$$\text{sin amplitude} = \text{sin dec sec lat}$$

This is proved with reference to Fig. 1.

Each of the diagrams in Fig. 1 represents the celestial sphere drawn on the plane of the horizon of an observer whose zenith is projected at Z. P represents the celestial pole and X a body rising having an amplitude of $\theta°$.

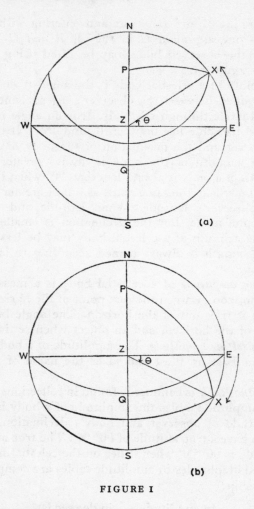

FIGURE I

In the quadrantal triangle PZX:

P is the hour angle of X at rising denoted by h

PX $= (90 \pm d)$, where d is the declination of X

PZ = (90 − ϕ), where ϕ is the observer's latitude

Z = (90 ± θ), where θ is the amplitude of X.

Applying Napier's rules to the triangle PZX, we have:

$$\sin \text{ co } PX = \cos Z \cos \text{ co } PZ$$

i.e.
$$\cos PX = \sin \theta \sin PZ$$

i.e.
$$\sin d = \sin \theta \cos \phi$$

and
$$\sin \text{ amplitude} = \sin \text{ dec sec lat}$$

Numerous azimuth diagrams have been devised to facilitate finding the azimuths of celestial bodies. An interesting azimuth diagram is that invented by a British Merchant Service master of the last century named Patrick Weir. Weir's diagram, which is still published by the British Hydrographic Department, and which until recently was popular in the Royal Navy, is a diagram on which hour angles and latitude are represented by confocal hyperbolae and ellipses respectively.

As well as graphical solutions for azimuths several graphical solutions for hour angle and zenith distance have been devised. An interesting diagram designed for this purpose is based on the stereographical projection of a model globe of radius six feet. The projection is divided into several sheets bound together and capable of giving zenith distances and hour angles to an accuracy of one minute of arc. This work was published by George Little-hales of the United States' Hydrographic Office.

The realization that latitude and local sidereal time (or local hour angle of the First Point of Aries) are defined by simultaneous altitudes of two celestial bodies of given declination and sidereal hour angle (or Right Ascension), led to the invention in 1923 of the *Two-Star Diagram* by K. Biej of the United States. Curves on the Biej diagram correspond to altitudes of selected stars. Simultaneous observations of the altitudes of the two stars enables the observer to find latitude and L.S.T. direct from the diagram. The longitude is then found by taking the difference between the L.S.T. and the G.S.T.

The renowned navigator of the present time, Captain, P. V. H. Weems of the United States, devised his famous *Star Altitude*

Curves on lines similar to those of Biej. Weems's *Star Altitude Curves*, which first appeared in 1928, are plotted on a chart on which the latitude scale conforms with that on a Mercator chart, so that the azimuth of the body to which a particular curve applies is at every point at right angles to the tangent to the curve at the point.

By means of Weems's *Star Altitude Curves* an observed position may be obtained from the altitude curves of the two or three selected stars in a matter of a couple of minutes without reference to declination, S.H.A., L.H.A., azimuth, D.R. position, and without the use of the *Nautical Almanac* or other tables with the exception of altitude correction tables.

The disadvantages of altitude curves are first, the limited number of stars for use with the method, and second, the curves apply to one epoch and it is necessary to apply corrections to the curves when used at a time different from that for which the curves are computed, to allow for the changing declination and S.H.A. of the stars employed, on account of precession and nutation of the Earth's axis, and the proper motion of the stars used.

The Baker navigation machine, invented in 1919 by Commander Baker, R.N., employs prepared altitude curves traced on a transparent tape wound on two rollers. The tape, which moves across a Mercator plotting chart, is marked with altitude curves for a series of suitable stars. Several tapes are provided, each occupying a small range of declination.

Numerous attempts have been made to provide the nautical astronomer with an instrument for the direct solution of the nautical astronomical problem. The globe of the early ocean navigators marked the first navigational instrument that could be adapted for this purpose. Although there have been many modern attempts to adapt a globe, on which is marked a graticule formed by the circles employed in nautical astronomy, for the purpose of solving PZX triangles with a high degree of accuracy, they have not met with commercial success. The term *star globe* applies to an instrument designed essentially to give approximate solutions to the problems of nautical astronomy.

The star globe of the present time is used chiefly for star identification purposes. A globe, usually about eight inches in diameter, is marked with the navigational stars, and with a graticule formed by parallels of declination and hour circles. It is housed in a box and may be rotated within a brass meridian ring which lies on a vertical plane, and which is graduated in degrees. The lower hemisphere of the globe lies within the box and is invisible. The top surface of the box within which the globe is housed is also graduated in degrees, and the intersection of this surface with the globe represents the observer's horizon.

The star globe is set by placing the celestial pole on the globe in such a position that its altitude is equal to the observer's latitude. By rotating the globe within the meridian ring until the hour circle corresponding to the observer's celestial meridian coincides with the meridian ring, the aspect of the visible hemisphere is truly portrayed for the time in question. The altitude and azimuth of a given star may then be lifted from the globe. Conversely using a given altitude and azimuth, a star may be identified.

A useful instrument which serves the same function as, but replaces the more costly star globe, is the *star identifier* or *planisphere*. A very popular star identifier is that invented by Captain G. T. Rude of the United States' Coast Guard. The *Rude Star Finder* consists of a thin white plastic disc about 8 inches in diameter and having a small pin at its centre. Navigational stars having north declination are marked on one side of the disc the centre of which represents the north celestial pole. The other side of the disc is marked with the navigational stars of the southern celestial hemisphere. The circumferential edge of this disc is graduated in degrees of L.H.A. ♈ from 0° to 359°. Together with the white opaque *base plate*, as the disc is called, is a series of transparent discs each having the same dimension as the base plate. Each of these may be fitted over the central pin of the base plate.

Each of nine transparent discs is marked with families of altitude and azimuth curves—the former ellipses and the latter hyperbolae—covering a range of 10° of latitude. A tenth transparent disc is printed in red with concentric circles representing parallels of declinations and intersecting diametrical lines representing hour circles.

To use the *Rude Star Finder*, the appropriate transparent disc is fitted over the base plate and oriented so that the north–south azimuth line on the transparent disc coincides with the L.H.A. ♈ on the base plate. The L.H.A. ♈ is found by applying the observer's longitude to the G.H.A. ♈ extracted from the *Nautical Almanac* for a given G.M.T. Having done this the star finder is correctly set for the purpose of finding the approximate azimuths and altitudes of the visible navigational stars. The red printed template is used for plotting the declinations and S.H.A's of the Moon and the navigational planets to enable the observer to estimate their altitude and azimuths at any given time.

The *Rude Star Finder* is a useful instrument for planning star sights for it enables an observer to ascertain the approximate altitudes and azimuths of the visible navigational stars. It may also be used to identify stars using their approximate altitudes and azimuths in order to find their declinations and S.H.A's.

The *Slide Rule* is sometimes used for solving nautical astronomical problems. The principle of the slide rule is that the logarithm of a product of two numbers is equal to the sum of the logarithms of the numbers. Thus, if two successive segments are set off along a straight line, of lengths equal to log A and log B on a given scale, their sum would be the log of the product of A and B on the same scale.

The object of an inventor of a navigational slide rule is to adapt the scales to suit the particular computations of nautical astronomy. Popular slide rules designed specifically for navigators are those invented by Podmore and Carmody.

The *Bygrave cylindrical slide rule*, invented in 1922 by Captain L. C. Bygrave, consists of two concentric cylinders which slide relative to one another. The inner cylinder is graduated with a spiral scale of log tangents, and the outer one with a spiral scale of log cosines. The solution of the PZX triangle by means of the *Bygrave slide rule* is based on a formula involving tangents and cosines. Two pointers are provided on a third sliding cylinder concentric with the other two. After a little practice the solution of altitude and azimuth may be found to an accuracy of a minute of arc within a couple of minutes of time.

PART IV

Practical Nautical Astronomy

CHAPTER I

The Use of the Nautical Almanac

The principal tables in the *Nautical Almanac* are found on the so-called *daily pages*. These tables facilitate finding for any given G.M.T. the G.H.A. and declination of any of the celestial bodies used in nautical astronomy.

The argument used in the tables of G.H.A. and declination is G.M.T., and the respondents are tabulated for every integral hour of G.M.T. for the whole year of the *Almanac*.

Interpolation for times other than integral hours of G.M.T. is facilitated by the interpolation tables described as 'Increment and Correction Tables', which are bound with the *Almanac*.

To find the G.H.A. and declination of the Sun at a G.M.T. other than one having an exact number of hours, the Sun table on the appropriate daily page is entered with the nearest G.M.T. preceding the given G.M.T. The tabulated G.H.A. and declination, and also the *d* value given at the foot of the table, are extracted. The Increments and Corrections Tables are then entered with the additional number of minutes and seconds as argument, and the increment from the column labelled SUN PLANETS is extracted. This, when added to the tabulated G.H.A., gives the required G.H.A. The correction to the tabulated declination is found from the same page as that used for finding the increment to the G.H.A., in this case entering the '*v* or *d* correction table' with *d* as argument. Care is necessary in applying the '*d* correction' in the correct sense. This is done by noting how the declination is changing: whether towards the north or south.

To find the G.H.A. and declination of the Moon for a G.M.T. other than one having an exact number of hours, the MOON table on the daily page is entered with the whole hour preceding the given G.M.T., and the tabulated values of the G.H.A. and declination are extracted. In addition, the corresponding 'v and d values' are lifted. The Increments and Corrections Table is then entered with the additional minutes and seconds of G.M.T. and the increment to the tabulated G.H.A. of the Moon is extracted from the column labelled MOON. The correction table on the same page is then entered with 'v' as argument and the 'v correction' extracted. Both increment and correction are then ADDED to the tabulated G.H.A. The procedure for finding the declination of the Moon is the same as that for finding the Sun's declination.

To find the G.H.A. and declination of a planet for any G.M.T. other than one having an exact number of hours, the procedure is similar to that for the Sun except that a 'v correction' is usually necessary. The 'v correction' is to be added for all planets with the possible exception for Venus. When finding the G.H.A. of Venus it is necessary to note whether the sign of the 'v value' on the daily page is negative or not.

To find the G.H.A. of a selected star, the appropriate daily page is entered with the integral hour of G.M.T. next preceding the given G.M.T. in the column labelled ARIES. The increments table is then entered with the additional minutes and seconds of G.M.T. and the increment to the tabulated G.H.A. of Aries is lifted. This is added to the tabulated G.H.A. of Aries to find the G.H.A. of Aries for the given G.M.T. The S.H.A. of the selected star is then extracted from the Star Table on the daily page. The S.H.A. of the star is added to the G.H.A. of Aries to give the required G.H.A. of the star. The declinations of the selected stars are tabulated in the Star Tables on the daily pages. Should the observed star not be a selected star it will be necessary to extract the star's S.H.A. and declination from the Navigational Stars Table to be found near the back of the *Almanac*.

Great care is necessary when using the daily pages of the *Almanac* to ensure that the required data for the correct day is extracted. The data for Aries, Moon and Sun for three consecutive days are arranged vertically on the daily pages, and it is a common blunder to extract the data from the wrong part of the page. A good practice, which helps to reduce this possibility, is to score out the data in the *Almanac* for each day as soon as convenient after the Greenwich date changes.

The worked examples given in the *Almanac* and the associated notes should be studied carefully.

RISING AND SETTING PHENOMENA

The times of Sunrise, Sunset, Moonrise and Moonset, and the times of the beginning and end of civil and nautical twilight, are given for the range of latitude between the parallels of 72° N. and 60° S., at a tabular interval of 2°, 5° or 10°. Tabulated times are given to the nearest minute, and special symbols are used to indicate that the Sun or Moon is circumpolar (□); twilight lasts all night (////); Sun or Moon does not rise above the horizon (▬). The tabulated times apply strictly to the middle of the three days of the page opening, and they are computed using the average values of declination and equation of time. For most cases, the tabulated times may be assumed to hold good for each of the three days. They are G.M.T's of the phenomena at the Greenwich meridian, and approximate to the L.M.T's of the phenomena over other meridians.

Interpolation for latitude and longitude, if necessary, is facilitated by using Tables I and II provided near the end of the *Almanac*.

The twilight tables are useful for planning star observations. The best time for star observations occurs when the horizon is still clear after sunset or before sunrise, and the required stars are still visible. This usually occurs near the time of the beginning of nautical twilight in the evening or the beginning of civil twilight in the morning.

THE CORRECTION OF SEXTANT ALTITUDES

The term sextant altitude applies to the angle read off the sextant after a celestial body has been observed for altitude. It is necessary to apply the altitude corrections described in Part II, Chapter II, to the sextant altitude to obtain the true altitude of the observed body.

Correction for index error (if any) is applied to the sextant altitude to obtain the *observed altitude*. The observed altitude may be described as the altitude read from the sextant when the instrument is in perfect adjustment. The remaining corrections are applied to the observed altitude to obtain the true altitude.

The practical method of correcting altitudes is to use the Altitude Correction Tables printed in the *Nautical Almanac*. These tables are located on the backs of the covers and on the fly-leaves. Those for the Sun, stars and planets, appear at the front of the Almanac, and those for the Moon at the back.

The dip correction is first to be applied to the observed altitude to obtain the *apparent altitude*. The apparent altitude is used as an argument when entering the appropriate Main Correction Table.

The Sun Altitude Correction Table provides for lower limb and upper limb observations. The corrections for lower limb observations are printed in heavy type and those for the less-frequently used upper limb observations are printed in light type.

To allow for the change in the Sun's semi-diameter the Sun Altitude Correction Table is in two parts. One part is for use during the period October to March, that is to say, the six-months period three months on each side of the approximate date of perihelion at which time the Sun's angular diameter is greatest. The other part is for use during the period April to September, that is to say, the six-months period three months on each side of the approximate date of aphelion, at which time the Sun's angular diameter is least for the year. It will be noticed that the difference between corresponding corrections for the two periods amounts to $0'\cdot2$.

The Sun Altitude Correction Tables are critical tables in which the variable interval of apparent altitude corresponds to a constant interval of $0'\cdot1$ in the corrections. If, for example, the apparent altitude of the Sun's lower limb during the period October to March is anything between $16°\ 28'$ and $15°\ 59'$, the main correction is $13'\cdot0+$. The tabular values are arranged so that at a critical entry the upper of the two possible values of the correction is to be taken. Thus, if the apparent altitude of the Sun's lower limb is $15°\ 59'$ the main correction is $12'\cdot9+$: if it is $16°\ 28'$ the main correction is $13'\cdot0+$.

The main altitude correction for Sun observations includes corrections for refraction, semi-diameter, parallax and irradiation.

The Altitude Correction Table for stars and planets is entered with the body's apparent altitude and the main correction is extracted. An additional correction is required for Venus and Mars to allow for parallax and phase. This additional correction varies with the time of year and with the altitude of the planet.

The Altitude Correction Tables for the Moon are entered with apparent altitude to obtain the first part of the correction. With the Moon's horizontal parallax (obtained from the daily pages), and the first part of the correction as arguments, the second part of the main correction is extracted.

Moon altitude corrections are always additive to the apparent altitude, but 30' is to be subtracted from altitudes of the Moon's upper limb. To facilitate correcting the Moon's altitude, a Dip Table is duplicated on the back flyleaf of the *Nautical Almanac*.

The Altitude Correction Tables described above are based on a mean refraction corresponding to that in which the air has a sea-level pressure and temperature of 30 inches of mercury and 50°F respectively. Should the atmospheric conditions of pressure and temperature be non-standard, it may be necessary to apply an additional correction to the main correction. This correction may be of particular significance when the altitude of the observed body is small.

The explanations of the Altitude Correction Tables that are given in the *Nautical Almanac*, and the examples provided, should be studied carefully by the student.

The Pole Star Tables provided in the *Nautical Almanac* are discussed in Part III, Chapter III.

An interesting and useful feature of the *Nautical Almanac* is the section headed PLANET NOTES. These notes are designed to assist in the identification of the navigational planets and Mercury, and to prevent confusing one with another. Associated with the Notes is an ingeniously contrived diagram from which the S.H.A. of any of the five planets Mercury, Venus, Mars, Jupiter and Saturn, and the local mean times of their meridian passages for any day of the year may be lifted. The navigator may see at a glance at the planet diagram which planets (if any) are suitably placed for morning or evening observation for any day of the year.

CHAPTER II

The Use of Sextant and Chronometer

To become an increasingly skilful sextant observer should be the aim of every nautical astronomer. It should be remembered that efficiency at observing with a sextant comes only after considerable experience. The novice should be warned to expect poor results from his sights, not through insufficient knowledge of principles but through lack of skill at using the principal measuring instrument of nautical astronomy.

The simplest sextant observation is that in which the noonday altitude of the Sun is measured. When the Sun is near meridian passage his altitude changes very slowly, thus facilitating the accurate measuring of his meridian altitude. In practice the altitude of the Sun's lower limb is measured. The procedure for so doing involves the following:

1. Ship the telescope and focus it. The common practice of marking the draw tube of the telescope to expedite setting the eyepiece for correct focusing is a good one: it often saves valuable time in cloudy weather.
2. Select the appropriate index and horizon shades. This technique involves trial and error which improves with practice.
3. Hold the sextant in the right hand with the arc of the instrument lying in the plane of the vertical circle through the Sun.
4. Observe the horizon through the unsilvered part of the horizon glass, and slide the index bar towards the observed object until the doubly reflected image of the object appears in line with the horizon.

5. Clamp the index bar, and use the tangent or micrometer screw to effect a grazing contact of the reflected Sun's lower limb with the true (direct) image of the horizon. To effect a grazing contact the least angle between the Sun's lower limb and the horizon must be measured. The sextant is rocked about a horizontal axis through the line of sight, so that the plane of the sextant arc sweeps through a small arc about the vertical circle through the observed body. The effect of this is for the reflected image of the observed body to sweep out an arc which, when the correct angle has been set by means of the tangent screw, just grazes the true image of the horizon.

It is important to remember that when taking a Sun-sight the Sun should never be observed directly through the sextant telescope unless the glass shades are in position. Temporary blindness or even injury to the eye may result if this rule is not obeyed.

If the sky is overcast it frequently happens that although the Sun is visible, his image through the sextant telescope is not clearly defined. In this circumstance the Sun is said to be *woolly*. When the Sun is woolly it is usually better to observe the altitude of his centre rather than that of his limb which latter is not nearly so well defined as his centre.

The Moon may be observed in the same way as the Sun, except that it may be necessary, on account of the phase of the Moon, to observe the upper instead of the lower limb. An alternative method of observing the Moon is the general method of taking star sights. Instead of the procedure outlined above, the index bar should be clamped to zero or near-zero on the arc, and the true and reflected images of the star should be observed simultaneously. By moving the index to higher readings on the arc, at the same time swinging the sextant in the vertical plane to keep the reflected image of the star in the silvered part of the horizon glass, the reflected image of the star is made to coincide with the true image of the horizon as observed through the unsilvered part of the horizon glass. When using this method, the observed object is said to be *brought down* to the horizon. Having brought down the observed object the tangent screw is used to effect a grazing contact to ensure that the arc of a vertical circle has been measured.

For star observations it is well to remember that by adjusting the rising piece of the sextant a greater or less amount of light enters the telescope from the horizon or from the observed body respectively. During the end of evening or the beginning of morning twilight, when the horizon is dim but the observed star bright, it is advisable to adjust the rising piece to ensure maximum amount of light entering the telescope through the unsilvered part of the horizon glass. Conversely, during the beginning of evening or the end of morning twilight, when the horizon is bright but the observed star dim, the rising piece should be adjusted to allow only a small amount of light entering the telescope through the unsilvered part of the horizon glass, so that the contrast between the brilliancy of the observed star and that of the horizon is sharp.

For measuring the altitude of a planet in daylight, the only satisfactory way is to compute the approximate altitude and azimuth of the planet for the time of the observation, and then to set the index on the sextant to the computed altitude on the arc. By holding the sextant with its arc lying in the vertical plane, and sweeping the horizon in the vicinity of the computed azimuth, the planet's image will be seen in the silvered part of the horizon glass provided that the computed altitude is within about a degree of the planet's actual altitude.

In misty or hazy weather, when the horizon is not clearly defined but celestial objects are visible, altitude observations are best made from the deck instead of from the bridge. By reducing the height of eye the distance of the sea horizon is also reduced. With a small height of eye the observer's horizon is nearer and therefore more clearly defined than when the height of eye is big.

When waves are running high the sea horizon will not be a sharp straight line as it is in clear weather with a calm sea. The line of sight of the sea horizon, assuming a stationary observer unaffected by the waves, is raised slightly by an amount which increases with the height of the waves. In practice, error due to this cause is ignored. When observing in rough weather it is best to observe from as high up as is conveniently possible. By so doing the sea horizon appears more nearly straight than it does when observed from nearer the sea surface.

In general a sextant observation must be timed by chronometer or stop-watch. An assistant, if available, should stand by the chronometer and at the instant when the altitude has been observed, the observer should shout 'time' or 'stop', whereupon the assistant should record the chronometer time of the observation.

When taking chronometer times the three hands of the instrument (second, minute and hour hands) should be read in order of their rapidity of motion. As well as recording the chronometer times of observations, the assistant should record the times by the chartroom clock to the nearest half minute, so that a check on the chronometer time-record is available.

If an assistant is not available the observer must time his own observations. The common practice is to count the seconds between the instant of observation and the instant when the chronometer time is noted, and then to reduce the chronometer time by the interval in seconds. The expert nautical astronomer should be able to count seconds accurately so that no material error results when timing his own sights.

If a watch is available, the altitude observations may be timed using the watch which should be held in the palm or strapped to the wrist of the left hand. As soon as the altitude has been measured the eye is shifted quickly from telescope to watch and the watch time recorded. This time should then be adjusted for the delay which should be ascertained from experiment and which should never be more than about a second.

The chronometer with which altitude observations are timed should be checked frequently (at least once per day when the ship is at sea) by Radio Time Signals. Full particulars of Radio Time Signals are given in the *Admiralty List of Radio Signals*, Volume 5.

There is an absence of uniformity in the systems used for transmitting radio time signals, but in 1955 the International Astronomical Union recommended the use of the method by which time signals controlled by the Royal Greenwich Observatory are transmitted. This method has become known as the

English System, and it is envisaged that in time it will replace all other methods.

In the English system time signals which are radiated for five minutes preceding each hour of G.M.T., consist of a series of 0·1 second dots at each second. The dots at the minutes are lengthened to 0·4 seconds to facilitate identification. The commencement of each dot is the timing reference point.

A widely used system of transmitting time signals is the International System known as ONOGO, the name being derived from the sequence of Morse letters used in the time code. The transmission takes three minutes, the procedure being:

1st minute: A series of the Morse letter X sent every five seconds from 0 to 49 seconds. This is followed by a six seconds period of silence followed by the Morse letter O (−−−) each dash of one second's duration commencing on the 55th, 57th and 59th second.

2nd minute: A series of the letter N (−.) sent once every 10 seconds commencing at the 8th, 18th, 28th, 38th and 48th second, the dot being given at every tenth second. This is followed by five seconds silence; followed by the Morse letter O (−−−) as in the preceding minute.

3rd minute: A series of the Morse letter G (−−.) sent once every 10 seconds commencing at the 6th, 16th, 26th, 36th and 46th second. This is followed by a five seconds silence; followed by the final signal the Morse letter O (−−−) as in the preceding minute.

Other systems in use are ONOGO (*Modified*), *United States' system*, *Modified Rhythmic*, and the *Russian Ordinary system*, each of which is described in the *Admiralty List of Radio Signals*.

The *Service Details of Radio Time Signals* include the name and call sign of the transmitting station and the radio frequency of the transmission, the system used, and the source of the time signal, the period of transmission, and perhaps other relevant information.

Many radio time signals are operated automatically by mechanism connected to the *Standard Clock* of an observatory. The accuracy of such signals is usually correct to within 0·05 second.

At some radio stations the time signals are sent by hand. The operator obtains the time from the standard clock at the radio station which is checked by astronomical observations or by reliable radio time signals. These signals are usually correct to within 0·25 seconds.

Of particular interest is the time signal transmitted by the B.B.C. This consists of the automatic transmission by the standard clock at the Greenwich observatory of six dots (or pips) representing successive seconds, the final dot being the time signal. This signal is usually accurate to within 0·1 second.

In some countries a telephonic time signal service is provided by the Post Office. In Great Britain oral announcements are made at intervals of 10 seconds by a speaking clock which is in operation in London and certain other centres. Full particulars of this service are given in the telephone directories for London and the other centres concerned.

CHAPTER III

The Navigational Astronomical Bodies

The navigational astronomical bodies include the Sun, Moon, the four planets Venus, Mars, Jupiter and Saturn, and the 173 stars for which astronomical data are given in the *Nautical Almanac*.

Throughout the hours of daylight, when the sky is clear and the horizon visible, the Sun is available for observation. For this reason he is often regarded as being the principal navigational body. Except on relatively rare occasions when the Moon or one of the navigational planets is visible during the daytime, the Sun alone provides the means of ascertaining a position line.

To find the ship's position during the daytime using the Sun alone, a running fix is necessary. This involves advancing or transferring the position line obtained from the first observation, through a course and distance corresponding to that made good by the ship during the interval between the times of the first and second observations. The running fix method is described in detail in Chapter V.

The running fix is less reliable than a fix obtained from simultaneous observations of celestial bodies. This follows because of the uncertainty in the assessment of the movement of the ship during the interval between the times of the first and second observations.

The optimum conditions for finding the ship's position using the running fix method in which two Sun observations are employed occur when the change in the azimuth of the Sun between the instants of the two observations is 90°, this angular

change taking place in the least possible time. The difference
between the Sun's azimuth at the times of the two observations
is equivalent to the angle between the two position lines obtained from the observations of the Sun at the two instants;
and, as we shall see in Chapter VI which deals with errors in
positions, the error due to any cause in a position obtained by
crossing two position lines is least when the position lines cross
at an angle of 90°.

By choosing the instants of observation such that the Sun's
bearing changes relatively rapidly and substantially in the
interval between the times of the observations, errors in the
estimated course and distance steamed during the interval are
kept as small as possible.

The common practice on merchant ships appears to involve
observing the Sun at about 8 a.m. and again at noon and finding
the ship's position by running fix for noon.

To get the most from the Sun the nautical astronomer should
have a clear understanding of the manner in which the Sun's
altitude and azimuth change during the day. These changes are
related to the Sun's declination, the observer's latitude, and the
time of day.

On the days of the equinoxes, when the Sun's declination is
zero, to an observer on the equator the Sun will rise bearing
due east at 6 a.m. and set bearing due west at 6 p.m., and will
cross the observer's meridian at his zenith. The Sun will change
his altitude at the uniform rate of 15′ per minute of time and
will not change his azimuth, except from due east to due west
at the instant he is at meridian passage. These conditions are
approximated for any observer in a low latitude on any day of
the year. In other words in low latitudes the Sun's rate of
change of altitude tends to be great and his rate of change of
azimuth tends to be small. In high latitudes, in contrast, the
Sun's rate of change of altitude tends to be variable and smaller
than it is in low latitudes, and his rate of change of azimuth
tends to be more uniform and greater than it is in low latitudes.
These matters are of great importance when planning Sun-
sights, not only for altitudes but for azimuths (in connection
with checking compasses) as well.

In low latitudes, because the change in the Sun's azimuth

during any given interval of time between two Sun-sights is relatively small, the running-fix method of finding the ship's position by Sun observations should be treated with caution, unless the course and distance made good during the interval between the two observations can be assessed accurately.

In low latitudes it may be possible and convenient to find the ship's position by means of two (or more) altitude observations of the Sun when he is near meridian passage using Captain Angus's method which is described in Part II, Chapter III.

Observations of the navigational planets are taken most commonly during twilight. Venus and Jupiter are sometimes well placed for observation during broad daylight. When this is so the astronomical navigator is provided with the means of crossing his Sun-sight position line. The Planet Notes and Diagram in the *Nautical Almanac* will assist the navigator who wishes to observe any of the navigational planets.

The Moon is often available for altitude observation during twilight, daytime, and even during the hours of darkness, provided that the horizon is distinctly visible in the direction of the bearing of the Moon. When observing the Moon at night at times when the sky is cloudy difficulty is often experienced in distinguishing the sea horizon from the horizontal edges of the long dark shadows of clouds that often appear on the sea surface in the direction of the Moon. When the sky is cloudless and the Moon's age is about 14 days, the brilliance of this luminary may be such that irradiation of the horizon immediately under the Moon may be considerable, this adding to the possibility of error in the true altitude. Apart from these factors the Moon is just as easy to observe as the Sun except that it may be necessary to observe the upper limb instead of the lower limb on account of the phase of the Moon.

When taking a Moon-sight it is important that the altitude of the upper limb or lower limb is observed, and not the altitude of a point on the Moon's *terminator*. The terminator is the line which separates the illuminated from the dark hemisphere of the Moon. On some occasions it is not obvious from the appearance of the Moon which of the upper or lower limbs is the illuminated limb. On these occasions the following particulars relating to the Moon should be considered when selecting the

Moon's limb. First, when the Moon's age is between 0 days and 14 days, that is to say, during the period between the times of New and Full Moon, the western limb of the Moon is illuminated. When the Moon's age is between 14 and 28 days her eastern limb is illuminated. The age of the Moon at any time may be ascertained from the *Nautical Almanac*, so that it is a simple matter to ascertain which is the illuminated side of the Moon. The second factor to bear in mind is that the straight line which joins the ends of the terminator is at right angles to the direction of the Sun from the Moon. The direction of the straight line which joins the ends of the terminator may be ascertained from a consideration of the relative positions of the Earth, Moon and Sun, and this in turn will enable the observer to select the Moon's limb.

Stellar observations have the advantage over Sun-sights in that simultaneous observations are possible, and by their means a ship's position may be found direct instead of by the running-fix method which is always necessary for Sun-sights.

Star-sights are facilitated by familiarity with the constellations. The ability to recognize stars instantly makes for speed in taking a series of star-sights. The nautical astronomer should aim to get his series of star observations in the shortest possible time so that the several sights of the series may be regarded as having been made simultaneously. An assistant employed to record the chronometer times of the observations is essential if the observer wants to ensure that his eyes will be tuned to the darkness throughout the period of the observations. The star observer who times his own observations inevitably loses time in 'getting his eyes' as seamen say, after they have been temporarily blinded by the relatively bright light of the chartroom. It is admitted that this trouble may be reduced by having orange instead of white lights in the chartroom.

A series of star-sights should be planned so that the resulting position lines cross at relatively large angles (never less than about 30°). The stars to observe are best chosen from a star identifier or star-globe set to the time at and the position from which the observations are to be made. The optimum time for observing should be ascertained after consulting the twilight tables in the *Nautical Almanac*.

When planning a series of star-sights, the rate of change of altitude of each of the stars to be observed due to a combination of the Earth's rotation and the movement of the ship over the ground, should be considered. The order of observation should agree with the order of the rates of change of altitude of the observed stars. The star whose altitude is changing most slowly should be observed first, and the one whose altitude is changing most rapidly should be the last to be observed. By so doing errors due to assuming the observations being simultaneously made are kept to a minimum.

The time of the last observation is usually taken to be the time of the combined sights. If the interval between the instants of the first and last observations of the series is unduly long (more than about five minutes) it will be necessary to advance or transfer the lines of position of all but that obtained from the last star observed to allow for the ship's movement between the time of the observation of the earlier star and that of the last star of the series.

Although the principal aim of the nautical astronomer is to obtain a ship's position by crossing two or more position lines, a single position line, besides having a potential value in that it may be transferred for use with another position line obtained from a later observation, often has direct value to a navigator.

If the direction of a position line differs from that of the ship's course line by a large angle, the position line provides useful information whereby the observer accurately may assess the ship's speed made good since the time of the last observed position. If, on the other hand, the direction of a position line and that of the ship's course line are the same or nearly so, a single position line may indicate the effect of current or wind across the course line.

A good position line running north–south provides the navigator with a reliable estimation of his ship's longitude, whereas one lying east–west gives a reliable estimation of the ship's latitude.

A transferred single position line may often be employed for fetching up harbour along a safe line of approach. This use of a single position line normally applies to terrestrial position lines but, on occasions, an astronomical position line may serve the same function.

18

CHAPTER IV

Direct Methods of Sight Reduction

Despite the availability of a profusion of short-method and inspection tables designed to facilitate the solution of the PZX triangle, a great number of navigators employ the long method of sight reduction, in which spherical trigonometry is used, to compute their astronomical triangles.

When using Sumner's modified method the angle P of the PZX triangle is computed using the three sides of the triangle. When using the intercept method the side ZX of the astronomical triangle is computed using the angle P and the other two sides. Let us consider some of the direct methods of computing P and ZX.

Given the three sides of an astronomical triangle, any of the three angles may be computed using the fundamental spherical cosine formula (see Appendix I). In the PZX triangle we have:

$$\cos P = \frac{\cos ZX - \cos PZ \cos PX}{\sin PZ \sin PX}$$

from which:

$$\cos P = \frac{\sin a \pm \sin l \sin d}{\cos l \cos d}$$

where a, l and d are the altitude of the observed body, the latitude used in the computation, and the declination of the observed body.

The solution of P using this formula is tedious because the formula is not suitable for logarithmic computation. This defect led to the invention of other formulae, most of which are derived from the cosine formula, but which have the advantage in that they are adapted for use with logarithms.

There is a singularly wide variety of methods for finding an angle in a spherical triangle using the three sides, and many of these methods have been used by navigators. A layman may well be astonished at the fact that seamen were not provided with a standard and universal method for solving their PZX triangle soon after there were PZX triangles to be solved. This was not the case until relatively recently.

The methods used at different times during the last two centuries for solving the PZX triangle have seldom provided the shortest, or the simplest, or even the most accurate solution. The method employed was often dependent upon which particular set of a multitude of available nautical tables a mariner was accustomed to use. Moreover, once a specified method had been accepted, mastered and committed to memory, a conservative seaman tended to use it throughout his sea-going days.

One of the earliest methods of solving angle P of the PZX triangle was invented by a French naval officer named Charles Borda. Borda's method is derived as follows:

Since $\cos P = 1 - 2 \sin^2 P/2$ we have in the PZX triangle:

$$1 - 2 \sin^2 P/2 = \frac{\cos ZX - \cos PZ \cos PX}{\sin PZ \sin PX}$$

$$= \frac{\sin a - \sin l \cos PX}{\cos l \sin PX}$$

and
$$2 \sin^2 P/2 = 1 - \frac{\sin a - \sin l \cos PX}{\cos l \sin PX}$$

$$= \frac{\cos l \sin PX + \sin l \cos PX - \sin a}{\cos l \sin PX}$$

$$= \frac{\sin (PX + l) - \sin a}{\cos l \sin PX}$$

and
$$\sin P/2 = \sqrt{\sec l \operatorname{cosec} PX \cos s \sin (s - a)}$$

where
$$s = \tfrac{1}{2}(PX + l + a)$$

Lieutenant Henry Raper, R.N., gave a modified and improved method for solving P based on Borda's method in his well-known *The Practice of Navigation*. Raper's modification involved using the table of log sine squares, which is what the present-day navigator would recognize as the log haversine table.

$$\text{Haversine } \theta = \tfrac{1}{2} \text{ versine } \theta = \tfrac{1}{2}(1 - \cos \theta)$$

Now

$$\cos \theta = 1 - 2 \sin^2 \theta/2$$

therefore:

$$\text{hav } \theta = \sin^2 \theta/2$$

so that in the PZX triangle:

$$\text{hav P} = \sec l \text{ cosec PX} \cos s \sin (s - a)$$

This formula is short and simple but suffers the disadvantage in that it requires the use of no less than five trigonometrical tables.

It is an easy matter to derive the so-called half-angle formulae for finding an angle in a spherical triangle using the three sides

$$\sin \text{P}/2 = \sqrt{\frac{\sin (s - \text{PZ}) \sin (s - \text{PX})}{\sin \text{PZ} \sin \text{PX}}}$$

$$\cos \text{P}/2 = \sqrt{\frac{\sin s \sin (s - \text{ZX})}{\sin \text{PZ} \sin \text{PX}}}$$

$$\tan \text{P}/2 = \sqrt{\frac{\sin (s - \text{PZ}) \sin (s - \text{PX})}{\sin s \sin (s - \text{ZX})}}$$

To which of these formulae, each of which is suitable for logarithmic computation, preference over the others ought to be given, should depend largely upon the value of P. It can be demonstrated that it is expedient to use the first or third when P is acute, and that the second is most suitable when P considerably exceeds 90°.

In times gone by the most commonly used half-angle formulae for solving the PZX triangle were those giving sin P/2 and cos P/2. The former was the favourite: seemingly because sines alone were employed in the solution. What has often been described as being the shortest direct method for solving P is that in which the formula for sin P/2 is modified for use with the haversine and sine tables.

Since hav $\theta = \sin^2 \theta/2$, it follows that:

$$\text{hav } P = \frac{\sin (s - PZ) \sin (s - PX)}{\sin PZ \sin PX}$$

i.e.
$$\text{hav } P = \sin (s - PZ) \sin (s - PX)$$
$$\times \text{cosec } PZ \text{ cosec } PX$$

The principal stumbling block in using any of the half-angle formulae for solving P was due to the difficulty the seaman had in handling the signs of the trigonometrical functions of angles in the second quadrant. This is probably one of the reasons why the versine and haversine have become popular amongst navigators.

The trigonometrical functions the versine and the haversine were adapted to nautical astronomical needs during the 17th and 18th centuries. Now versine $\theta = 1 - \cos \theta$, so that the spherical cosine formula may be reduced to:

$$\text{vers } P = \frac{\text{vers } ZX - \text{vers } (PZ \sim PX)}{\sin PZ \sin PX}$$

Now,

$$\sin PZ \sin PX = \tfrac{1}{2}\{\cos (PZ \sim PX) - \cos (PZ + PX)\}$$
$$= \tfrac{1}{2}[\{1 - \cos (PZ - PX)\}$$
$$- \{1 - \cos (PZ \sim PX)\}]$$
$$= \text{hav } (PZ + PX) - \text{hav } (PZ \sim PX)$$

Also,

$$\tfrac{1}{2} \text{vers } P = \frac{\tfrac{1}{2} \text{vers } ZX - \tfrac{1}{2} \text{vers } (PZ \sim PX)}{\sin PZ \sin PX}$$

therefore:

$$\text{hav } P = \frac{\text{hav } ZX - \text{hav } (PZ \sim PX)}{\text{hav } (PZ + PX) - \text{hav } (PZ \sim PX)}$$

This formula is called the all-haversine formula. Its great advantage is that the only trigonometrical function involved in its use is the haversine.

The significant feature of the versine (and haversine) is that it has a unique positive value for every angle between 0° and 180°. It follows that angles in the second quadrant present no difficulty in respect of algebraic sign when the versine or haversine is used instead of the fundamental trigonometrical functions.

We have noted that the cosine formula for angle P in the PZX triangle adapted for use with versines is:

$$\text{vers } P = \frac{\text{vers ZX} - \text{vers (PZ} \sim \text{PX)}}{\sin \text{PZ} \sin \text{PX}}$$

i.e. $$\text{vers } P = \frac{\text{vers } z - \text{vers } (l \pm d)}{\cos l \cos d}$$

i.e. $$\text{vers } P = \{\text{vers } z - \text{vers } (l \pm d)\} \sec l \sec d$$

i.e. $$\text{vers } P = \text{vers } \theta \sec l \sec d$$

also $$\text{hav } P = \text{hav } \theta \sec l \sec d$$

where $$\text{hav } \theta = \text{hav } z - \text{hav } (l \pm d)$$

This is the direct method for finding angle P of the PZX triangle, most frequently used by navigators of the British Merchant Navy.

The haversine formula, as the above formula is called, may be transposed thus:

$$\text{hav } z = \text{hav } P \cos l \cos d + \text{hav } (l \pm d)$$

In this form it is used when computing the zenith distance in order to find an intercept. The formula in this form is usually known as the cosine–haversine formula. It was introduced by Percy L. H. Davis of the British Nautical Almanac Office in his *Requisite Tables* first published in 1905.

Davis, in his *Requisite Tables*, was first to publish a haversine table giving both natural and logarithmic values side by side in a common table in order to facilitate the solution of the astronomical triangle using the method he introduced.

An interesting direct method, similar to Davis' cosine–haversine method, was introduced by the Japanese naval officer S. Yonemura. Yonemura's method was published in Ogura's tables in 1920.

In the cosine–haversine method for finding ZX in the PZX triangle, viz.:

$$\text{hav ZX} = \text{hav} (l \pm d) + \cos l \cos d \text{ hav P}$$

let $\cos l \cos d \text{ hav P} = \text{hav } \theta$. Then

$$\text{hav ZX} = \text{hav} (l \pm d) + \text{hav } \theta$$

Now,

$$\frac{1}{\text{hav } \theta} = \sec l \sec d \frac{1}{\text{hav P}}$$

This is Yonemura's formula the solution to which is facilitated by means of a table giving the logs of reciprocals of haversines.

In solving a sight by direct method, a latitude in the case of the modified Sumner's method, or a latitude and longitude in the case of the intercept method, must be used in the computation. In general, therefore, a position is used in working out a sight. This position is often described loosely as the *ship's position*. It is simply a position used in the computation and as such it may best be described as a USED *position*. The used position must, of necessity, be not too distant from the actual but unknown position of the ship at the time of the observation. In many cases it is best to use a position the latitude of which is an exact number of degrees. This facilitates the use of the log table when extracting the secant or cosine of the latitude.

The used position should be based on the ship's estimated position (E.P.). The E.P. of a ship at any time is derived from the ship's dead reckoning (D.R.) position for the same time.

The derivation of the term *dead reckoning* is not known with certainty. The term in present usage denotes a position obtained by applying the course and distance made by the ship through the water to the last known observed position or fix. It follows that the ship is seldom at her D.R. position at any given time for the simple reason that the course and distance made through the water is seldom the same as that made over the ground. The effect of wind, current, bad steering, the heave of the sea, is generally to cause the ship's actual position at any time to be different from her D.R. position at the given time.

The best estimation of the ship's position for any time is

made by applying to the ship's D.R. position an estimation of the effects of each of the above causes (wind, current, etc.). The resulting position is called the *estimated position* (E.P.).

The ability to derive a good E.P. is perhaps the hallmark of the expert navigator. Good judgement, gained through experience, of the effects of the factors influencing the way of his ship, is necessary in obtaining a good E.P.

CHAPTER V

The Daily Routine of the Nautical Astronomer

The master of a merchant ship carries the burden of responsibility for the safety of the ship under his command. Related closely to safety is the navigation of the ship. To know where the ship is at any time and, more important, to know how the ship is moving over the ground and to be able to forecast her likely position at any time in the near future, are the principal problems of the navigator. When the ship is away from land, in the absence of electronic aids to navigation, the principles and practice of nautical astronomy must be brought to bear in seeking answers to these problems.

In the discussion on nautical astronomy presented in the foregoing pages we have said little about charts and compasses. These are amongst the more important of the instruments of navigation. Their care and management, in common with most other navigational equipment, falls to the charge of the navigating officer of the ship. The Second Mate in a merchant ship is usually designated the navigating officer, and it is this officer who normally is responsible to the ship's master for seeing to it that the charts and compasses and all other navigational equipment is looked after properly and that it is available for instant use when required.

The navigation of a ship is collectively performed by the officers of the watch. Each officer during his watch sees to it that the ship is never on an unsafe course. He checks the ship's rate of progress, using compass and patent log and, when possible and practicable, by finding her position by observation. In this brief chapter we shall discuss the normal routine in the navigator's day's work, when his ship is at sea away from the land.

Of great importance is the record that should be kept of courses steered and distances made through the water on each course, together with other information of relevance to navigation. This record is the *logbook* which is brought up to date by each officer at the end of his watch. It is from the logbook record that the ship's D.R. position and Estimated Position for any time may be found. Information of navigational importance, in addition to courses and distances, that should be recorded, includes: the direction and speed of the wind, the set and rate of the current, and alterations of course and speed made in order to keep clear of other vessels.

The charts for use during a voyage should be corrected to the date of the latest available *Notice to Mariners*, and they should be arranged sequentially in order of use in a chartroom drawer. The chart used when the ship is away from the land is usually a small scale general chart of the area on or from which rough courses and distances may be plotted or measured.

The *Sailing Directions* of the area in which the ship is being navigated should have been studied before the commencement of the voyage, and they should be available for use at every stage of the voyage.

The chronometers should be wound at the same time by the same officer each morning. The chronometer error should be found from radio time signals at least once a day when the ship is at sea and a record kept, in the chronometer journal, of the daily rate and the accumulated error.

The compass bearing of the Sun during the daytime, or that of a star at night, should be observed at least once during each watch, and compared with the body's true bearing to obtain the compass error for the heading of the ship at the time of the observation.

The true bearing of a celestial body, when it is needed for compass checking, is usually lifted from azimuth tables such as those of Davis or Burdwood, or from *A, B and C Tables*. Care should be taken to interpolate properly when using these tables, especially when the observed body is changing its azimuth rapidly. Celestial objects are most suitable for observing for azimuth when their altitudes are small and their rates of change of azimuth are not great.

Some navigators employ amplitude tables when checking compasses. These tables are worked out for an object whose true zenith distance is 90°. In other words, the amplitudes extracted from the table, using as arguments latitude of observer and declination of observed body, apply to a celestial body the true altitude of whose centre is 00° 00′.

When using the Moon or Sun for amplitude, care should be taken to ensure that the centre of the body lies on the observer's celestial horizon. At the instant when this is so, the position of the object's centre relative to the observer's visible horizon will vary according to the body observed, and the height of eye of the observer.

In high latitudes the diurnal circles of celestial objects which rise and set cross the horizon at a very small angle. It follows that in these circumstances, a small change in altitude results in a relatively large change in the body's azimuth. For this reason amplitude observations are particularly liable to error in high latitudes. The navigator is advised to ignore amplitude tables and to treat every observation of a celestial body for checking compasses as an azimuth observation, timing the observation to obtain the body's hour angle, and using the azimuth tables to ascertain the body's true azimuth.

Astronomical sights taken when the ship is at sea should include twilight observations of stars during both morning and evening. In planning star sights the times of Sunrise and Sunset and the times of morning and evening twilight should be obtained from the *Nautical Almanac*.

Because of the uncertain effects of astronomical refraction at low altitudes, stars chosen for altitude observation should have altitudes of more than about 10°.

Stars whose altitudes are small (less than about 15°) usually change their altitudes relatively rapidly. Altitude observations in these circumstances, therefore, are attended with difficulty. On the other hand a star whose altitude is large (more than about 60°), is not easy to observe on account of the small degree of curvature of the grazing arc of its reflected image when the sextant is rocked during the observation.

The nautical astronomer should have no difficulty in recognizing stars observed during morning twilight, as he will have had the opportunity of studying them during the darkness before the time of sights. Stars observed during evening twilight are not so readily identified at the time of sights, and may have to be identified later.

During evening twilight the eastern part of the sky becomes darker more quickly than the western part. Conversely, during morning twilight the eastern part of the sky becomes brighter more quickly than the western part. For this reason it is best that bodies in the eastern sky during morning twilight, and those in the western sky during evening twilight should, in general, be observed before those in the opposite half of the sky.

During the daytime the Sun is available for fixing by the running-fix method. It is customary to find a position for noon each day by crossing position lines obtained from an a.m. observation and a meridian altitude observation of the Sun respectively.

During daytime the Moon, when in the first quarter during the afternoon, or the last quarter during the forenoon, is often suitably placed for simultaneous observation with the Sun. Occasionally Venus and/or Jupiter is available for daytime observation, and these bodies on these occasions provide the means of fixing by simultaneous observations with the Sun.

Mainly for record purposes the course and distance made good between successive noons is worked out using the so-called observed positions for the two noons.

The working out of the noon position by observation, employing the running-fix principle, is normally a process of computation in which the principles of position-line navigation are lost to sight and mind.

The morning sight is usually worked out using the ship's D.R. latitude at the time of the observation, and a longitude is computed. The d.lat and d.long corresponding to the course and distance made good between the times of morning and noon sights, are applied to the used latitude and the computed longitude respectively to give a position which is the ship's noon position only if the latitude by meridian altitude observation is the same as that obtained by applying the d.lat to the latitude used in computing the morning sight. If the noon latitude by observation is different from the latitude run up to noon it will generally be necessary to apply a longitude correction to the longitude obtained by applying the d.long to the computed longitude from the a.m. sight. The longitude correction is obtained from the *A, B and C Tables* and is based on the relationship:

Error in longitude = Error in latitude × C correction

The longitude correction is usually applied by a rule of thumb, in which the Sun's azimuth at the time of the morning sight figures, and the so-called observed longitude, is obtained as if by magic.

An alternative method of finding the noon position using the running-fix method is to use a *plotting sheet* on which the position lines associated with the sights are drawn, thus manifesting the basic principle of the problem, and at the same time giving an indication (from the angle of cut of the position lines) of the degree of accuracy and reliability of the position obtained.

Plotting should be regarded as being an important part of the work of an astronomical navigator. The tools of plotting include a plotting sheet (usually a page of the navigator's workbook), a sharp pencil, a graduated straightedge, a pair of dividers, and a good protractor. For plotting on a navigational chart, a parallel rule is almost a necessity.

All significant lines and points on a plotting sheet or chart should be labelled or indicated in a conventional manner. It is customary to use the following symbols for this purpose:

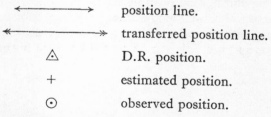

Course lines should be labelled with the true course printed neatly along the line in three-figure notation. Times should be printed in 24-hour notation. All inscriptions should be neat and legible and they should be located so that they least interfere with the navigator when he uses the chart or plotting sheet.

When plotting on a navigational chart it is important that the chart in use alone occupies the chartroom table. The practice of using one chart placed on top of another should be condemned, on account of the possibility of using the scale of the chart below for marking off distances and positions on the chart on top. Seldom do navigational charts have the same scales.

For the small-scale plotting of position lines obtained from simultaneous star sights, most navigators use their work books. In this case the scale used for plotting is usually 1 inch to 10 miles, a scale too small for high accuracy. For preference plotting should be done on a relatively large scale (1 cm to a mile is suitable). The back of a cancelled chart on which a large compass rose is centrally drawn provides an ideal plotting sheet.

When using a plotting sheet for simultaneous star-sights, or even running-fix Sun-sights, the intercept method is to be preferred to the modified Sumner method. The plotting sheet, however, does not lend itself for use with the a.m. Sun-sight run up to noon: and this, no doubt, is a major reason why the intercept method is not universally used for Sun-sights.

It is important that the navigator works accurately and methodically when solving his sights. To facilitate accurate working it is advisable to keep to a standardized system of solving sights. Most navigators use systems which they have evolved according to their varied experiences and sight-working habits. Familiarity with the layout of a solution to a sight or series of sights assists in checking and finding possible mistakes.

The practical as well as the theoretical aspect of dealing with errors in positions and position lines, and a discussion on the cocked hat and multiple star fixes, will be given in the following chapter.

CHAPTER VI

The Treatment of Navigational Errors

In this chapter we shall discuss the errors in position lines and positions obtained from astronomical observations due to various causes.

ERROR IN A POSITION LINE DUE TO ERROR IN ALTITUDE

An error in an observed altitude due to any cause displaces the resulting position line by an amount equivalent to the error at the rate of 1 mile per minute of arc error. The effect on an intercept resulting from an error in altitude is dependent upon the sense of the error and the name of the intercept. If the incorrect altitude is too large the effect of the error is to displace the position line in the direction of the azimuth of the observed body. If the incorrect altitude is too small the effect is to displace the position line in the opposite direction to that of the azimuth of the observed body.

An error in an altitude may result from not applying index error properly, from incorrect application of altitude corrections, from using incorrect values of altitude corrections, especially those of dip and refraction, or from the observer's personal equation.

The error in latitude and longitude resulting from an error in altitude is investigated with reference to Fig. 1.

In Fig. 1, C represents a ship's actual position. CL represents part of the parallel of latitude through C, and CM represents part of the meridian through C. Let the azimuth of the observed body *X be Z.

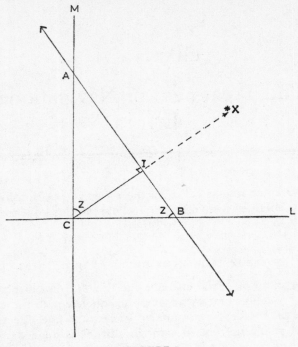

FIGURE I

Let us assume that an observation of the body *X, using the ship's actual position in the computation of the PZX triangle, yields an intercept equivalent to CI where I is the intercept terminal position. Had there been no error in altitude the intercept would have been zero and C and I would have been coincident. The computed intercept, therefore, is equivalent to the error in the altitude.

The resulting error in latitude is equivalent to AC.

From the triangle ACI:

$$AC = CI \sec ACI$$

therefore:

$$\text{Error in latitude} = \text{Error in altitude} \times \sec Z$$

The resulting error in departure is equivalent to BC.

From the triangle CIB:

$$CB = CI \text{ cosec } IBC$$

therefore:

$$\text{Error in dep} = \text{Error in altitude} \times \text{cosec } Z$$

Now dep = d.long cos lat (parallel sailing formula), therefore:

$$\text{Error in longitude} = \text{Error in altitude} \times \text{cosec } Z \text{ sec } \phi$$

where ϕ is the latitude of the observer.

ERROR IN A POSITION LINE DUE TO AN ERROR IN TIME

An error in G.M.T. will result in an error in hour angle, This, in turn, will lead to an error in the computed zenith distance

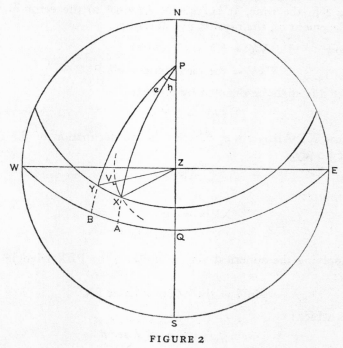

FIGURE 2

when using the intercept method of sight reduction. This will cause an error in the computed intercept, this resulting in a displacement of, or error in, the position line. The effect of an error in G.M.T. on an intercept is investigated in Fig. 2.

Fig. 2 illustrates the visible celestial hemisphere drawn on the plane of the horizon of an observer whose latitude is ϕ and whose zenith is projected at Z. P is the projection of the elevated celestial pole; N, E, S and W are the projections of the cardinal points of the horizon and WQE is the projection of the equinoctial.

Let the hour angle of the observed body (which is projected at X in Fig. 2) be h. Let the error in the G.M.T. be denoted by e, which is represented by angle XPY in Fig. 2.

Let the point V lie on YZ such that ZV = ZX. The triangle XVY, being small, may be regarded as being a plane triangle right-angled at V.

Arc VY is the error in zenith distance (and altitude) due to error e in the time. It is, therefore, equal to the error in, or displacement of, the resulting position line.

Now $VY = XY \sin YXV$

but $YXV = $ the parallactic angle PXZ

Let this angle be denoted by θ. Then

$$VY = XY \sin \theta$$

Now $XY/AB = \cos d$, where d is the declination of the observed body.

Now $AB = e$

Therefore:

$$XY = e \cos d$$

and $$VY = e \cos d \sin \theta$$

Applying the spherical sine formula to the PZX triangle we have:

$$\sin \theta = \sin PZ \sin Z \operatorname{cosec} PX$$

from which:

$$\sin \theta = \cos \phi \sin Z \sec d$$

It follows that:

$$VY = e \cos d \cos \phi \sin Z \sec d$$

i.e.

$$VY = e \cos \phi \sin Z$$

therefore:

$$\frac{\text{Error in altitude}}{\text{(in min of arc)}} = \frac{\text{Error in time}}{\text{(in min of arc)}} \times \cos \phi \sin Z$$

$$\frac{\text{Error in altitude}}{\text{(in min of arc)}} = \frac{\text{Error in time}}{\text{(in sec of time)}} \times \frac{\cos \phi \sin Z}{4}$$

From this formula it may readily be shown that error in altitude due to error in time is zero when $\cos \phi$ or $\sin Z$ is zero. In other words in latitude 90°, regardless of the azimuth (which incidentally is always 180° in latitude 90° N., and always 000° in latitude 90° S.) any error in time will cause no error in computed altitude (or zenith distance). Also, in any latitude, when an observed body bears 000° or 180°, any error in time will cause no error in computed altitude (or zenith distance).

Error in computed altitude (or zenith distance) due to error in time is greatest for any given latitude when $\sin Z$ is maximum. When an observed body is on the prime vertical circle of an observer: that is to say, when its azimuth is 090° or 270°, error in altitude (or zenith distance) due to error in time is greatest.

ERRORS IN RUNNING FIXES

a. Error in Transferred Position Line due to an Error in the Distance

The displacement in a transferred position line due to an error in the distance run between the times of the sights is investigated with reference to Fig. 3.

Let AB in Fig. 3 be a position line obtained from a celestial observation. Let CX be the true distance run between the times of observations from which a running fix is obtained. Let the error in the distance be XX_1, so that the false transferred position line is $A_F B_F$, and the true transferred position line is $A_T B_T$. The error in the transferred position line is XY which is denoted by e in Fig. 3.

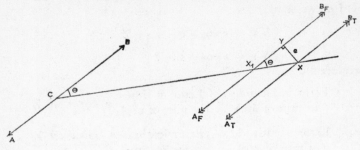

FIGURE 3

In the triangle XYX_1

$$e = XX_1 \sin \theta$$

where θ is the angle between the position line and the course line.

The formula shows that the error in a transferred position line is maximum when $\sin \theta$ is maximum: that is, when θ is 90°.

When θ is 0° error in a transferred position line due to an error in the distance run is zero. It follows that an error in a transferred position line is zero if the observed body lies abeam to port or starboard, and it is greatest when the observed body lies dead ahead or right astern at the time of the first observation.

b. Error in Transferred Position Line due to an Error in the Course

The error in a transferred position line due to an error in the course made good during the interval between two sights used for a running fix, is investigated with reference to Fig. 4.

Let AB in Fig. 4 represent a position line obtained from an observation of a celestial body *X whose azimuth is N. $(90 - \phi)$ W. Let the distance run on a course N. θ E. between the times of the observations be d, so that the true transferred position line is $A_T B_T$. Let the error in the course be α, so that the false transferred position line is $A_F B_F$, and the corresponding error in the transferred position line is e.

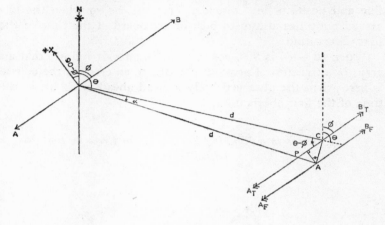

FIGURE 4

In the right-angled triangle APC:

$$AP = e$$

$$PAC = (\theta - \phi)$$

$$\therefore e = AC \cos (\theta - \phi)$$

If α is a small angle and AC is regarded as being very small compared with d, then:

$$AC = d \sin \alpha \quad \text{(very nearly)}$$

or $$AC = d\alpha \text{ radians}$$

Therefore: $$AP = d\alpha \cos (\theta - \phi)$$

or $$e = \frac{d\alpha^\circ \cos (\theta - \phi)}{57.3}$$

When $\theta = \phi$, $(\theta - \phi)$ is $0°$ and $\cos (\theta - \phi)$ is unity and maximum. Thus, for any given distance and error in course, the error in the transferred position line is greatest when the course angle is equal to the angle which the position line makes with the meridian. In other words, error in a transferred position line due to an error in the course is a maximum when the course line and position line coincide. This will be so when the observed body lies abeam to port or starboard at the time of the first observation.

When $(\theta - \phi)$ is $90°$, $\cos(\theta - \phi)$ is zero. It follows that an error in a transferred position line due to an error in the course is zero when the observed body is dead ahead or astern at the time of the first observation.

c. Error in a Running Fix due to an Error in the Distance Run

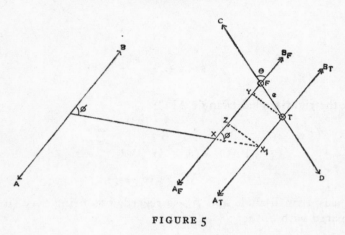

FIGURE 5

In Fig. 5, AB represents the first position line and CD represents the second position line with which the first position line transferred is crossed to produce a false running fix at F.

Let the angle between the first position line and the course line be ϕ, and let the error in the distance run be XX_1, in which case the false transferred position line is $A_F B_F$, and the true transferred position line is $A_T B_T$. The ship's true position at the time of the second sight is T, and the error in the fix is FT, denoted by e.

Let the smaller angle between the two position lines be θ.

In the triangle TYF:

$$e = YT \operatorname{cosec} \theta$$

In the triangle XZX_1:

$$ZX_1 = XX_1 \sin \phi$$

But $$ZX_1 = YT$$
therefore:

$$e = XX_1 \sin \phi \operatorname{cosec} \theta$$

Examination of this formula reveals that the error in a running fix due to an error in the run is zero when $\sin \phi$ is zero: that is, when ϕ is 0°. This occurs when the transferred position line coincides with the original position line, and this is so when the ship's course and the first position line coincide.

The error in a running fix due to a given error increases for any given value of ϕ as the angle of cut θ decreases. The formula shows that an optimum condition for a running fix occurs when the angle of cut is 90°, in which case $\operatorname{cosec} \theta$ is unity.

d. *Error in a Running Fix due to an Error in the Course*

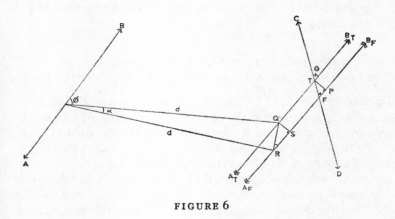

FIGURE 6

In Fig. 6, AB represents the first-position line and CD represents the second-position line obtained after the ship had run *d* miles from the time of the first observation. Let the error in the course be $\alpha°$ as shown in Fig. 6. The false transferred position line is $A_F B_F$ and the true transferred position line is $A_T B_T$.

T and F represent the true and false running fixes respectively. Let the angle of cut between the first- and second-position lines be θ, and let the angle between the first-position line and the course line be ϕ.

The error in the transferred position line is SQ, which is equal to PT. The error in the running fix is, therefore:

$$e = \text{PT cosec } \theta$$

But
$$\text{PT} = \frac{d\alpha \cos \phi}{57 \cdot 3} = \text{SQ}$$

therefore:

$$e = \frac{d\alpha \cos \phi \text{ cosec } \theta}{57 \cdot 3}$$

This formula shows that when $\cos \phi$ is zero, that is, when ϕ is 90°, the error in a running fix due to an error in the course is also zero. This applies when the object observed for the first-position line is right ahead or right astern at the time of the first observation.

ERROR IN LATITUDE DUE TO TREATING THE MAXIMUM ALTITUDE AS THE MERIDIAN ALTITUDE

We have seen in Part II, Chapter V, that the observer's northerly or southerly motion and the changing declination of a heavenly body influence the time at which the body reaches its maximum daily altitude. We have also seen that if the combination of the rates of change of the observer's latitude and the body's declination is towards the geographical position of the body when it is near meridian passage, the time of maximum altitude occurs after the time of meridian altitude. If away from the geographical position of the body the time of maximum altitude is earlier than the time of meridian altitude. In the same chapter a formula for giving the interval between the times of maximum and meridian altitudes is investigated. It is now our aim to consider the difference between the maximum and meridian altitudes of a heavenly body. This difference is equivalent to the error in a latitude obtained from an observation of a body on the meridian when its maximum instead of its meridian altitude is used.

The northerly or southerly rate of motion of a ship may readily be found from the Traverse Tables, this rate being equal to the product of the ship's speed and the cosine of the course angle. The rate of change of a body's declination may be found by inspection from the *Nautical Almanac*. Let the combination of these rates be y' per hour.

If the interval between the times of maximum and meridian altitudes is computed it is an easy matter to estimate the difference between maximum and meridian altitudes if y is known.

If, for example, the interval between the times of maximum and meridian altitudes is say 10 minutes, and y is say 18′ per hour, the difference between maximum and meridian altitudes will be $\frac{10}{60} \times 18$, that is 3′ approximately. The solution is approximate because we have assumed that the rate of change of altitude at the time of maximum altitude is uniform. This is a false assumption, but the error introduced when estimating the difference between maximum and meridian altitudes in this way is trifling.

On fast ships, especially when steaming along or nearly parallel to a meridian, and especially at or near the time of an equinox, when the Sun's declination is changing most rapidly, it is important when using the noonday Sun for latitude to compute the time of meridian passage and to take the sight at this pre-computed time.

ERROR IN FIX OBTAINED FROM SIMULTANEOUS OBSERVATIONS

The minimum amount of information necessary for fixing a ship from astronomical observations consists of two intersecting position lines. A position obtained in this way is sometimes called a *two-star fix*. If more than two stars are observed in order to find the ship's position, the position is referred to as a *three-star fix* when three position lines intersect; a *four-star fix* when four position lines intersect, and a *multi-star fix* when more than four stars are observed. Let us deal with errors that may occur in such fixes.

a. The two-star fix

An interesting and fruitful way of dealing with errors in positions obtained from astronomical observations involves a consideration of the bisectors of pairs of position lines.

The use of bisectors for analysing star sights was brought to the notice of navigators by Admiral L. Tonta of the Royal Italian Navy in 1931 in an article in *Hydrographic Review Vol. 8*, and more recently by Captain Mario Bini, of the Italian Navy, in a valuable paper which appears in Volume 8 (1955) of the *Journal of the Institute of Navigation*.

Any pair of intersecting straight lines produces two bisectors mutually perpendicular to one another. The bisector of any pair of intersecting astronomical position lines with which we shall be concerned is that bisector which not only bisects the position lines but also bisects the angle contained between the directions of the observed bodies which produce the position lines.

Fig. 7 illustrates two astronomical position lines obtained from simultaneous observations of celestial bodies whose azimuths at the time of the observations are indicated by the arrows labelled *X and *Y.

In Fig. 7, BB₁, the bisector which bisects the angle XOY also bisects the position lines.

Should the azimuths of two observed bodies differ by 180°, and the position lines obtained from the observations not be

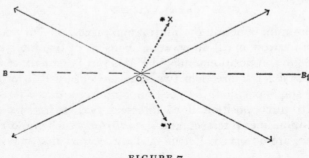

FIGURE 7

coincident, the bisector of the position lines is taken to be the line which lies midway between the two position lines.

Consider the two position lines illustrated in Fig. 8. These position lines intersect at F which, if there is error in one or both of the position lines, is a false fix.

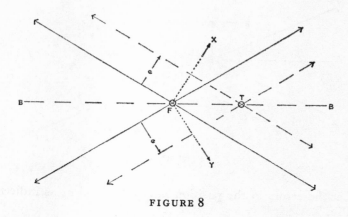

FIGURE 8

Let us suppose that the position lines illustrated in Fig. 8 have been plotted with the same systematic error—an error perhaps due to the application of the wrong index error, or a constant personal error—affecting the two position lines. Let this systematic error result in an error in each of the position lines amounting to e, as illustrated. The true position of the ship is, therefore, at T.

It is easy to see that, regardless of the magnitude or the sense of e, the true position of the ship lies on the bisector of the two position lines, provided that the same systematic error, and no random error, affects both position lines.

It must be emphasized that the true position of the ship lies on the bisector of two position lines only if systematic error alone influences the observations.

Let us now consider a case in which two position lines obtained from astronomical observations are each affected by a combination of systematic and random errors. Referring to Fig. 9, suppose that simultaneous observation of two stars *X and *Y result in two position lines XX_1 and YY_1 intersecting at F.

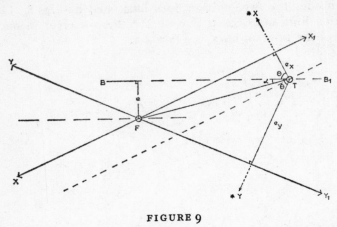

FIGURE 9

Let the errors in the position lines be e_x and e_y as indicated in Fig. 9.

Let T in Fig. 9 represent the ship's true position. Then BB_1 is the bisector that would have been obtained had the position lines been free from error. Let the displacement of the false fix F from this bisector be e. This quantity e may be regarded as being the error affecting the bisector drawn through the false position F.

Let the smaller angle between the directions of the observed bodies be 2θ. The angle between the direction of either body and that of the bisector is, therefore, θ. Let the angle which FT makes with the bisector be α, then:

$$e_x = FT \cos (\theta + \alpha) \tag{1}$$

$$e_y = FT \cos (\theta - \alpha) \tag{2}$$

$$FT = e \operatorname{cosec} \alpha \tag{3}$$

From (1) and (2) we get:

$$e_x = FT(\cos \theta \cos \alpha - \sin \theta \sin \alpha) \tag{4}$$

$$e_y = FT(\cos \theta \cos \alpha + \sin \theta \sin \alpha) \tag{5}$$

Subtracting (4) from (5) and substituting for FT from (3) we have:

$$e_y - e_x = 2e \sin \theta$$

from which:

$$e = \frac{e_y - e_x}{2 \sin \theta}$$

It may readily be seen from this formula that for any given errors in the position lines, the error in the bisector is least when $2 \sin \theta$ is maximum: that is to say, when $\sin \theta$ is 1. This is so when the azimuths of the two observed bodies differ by 180°. In this circumstance the condition for using a bisector is optimum.

Considering the manner in which the sine of an angle changes as the angle changes we may regard the optimum condition to pertain so long as θ is greater than about 70°. This is to say when the azimuths differ by more than about 140°. When θ is less than about 30°: that is to say, when the difference between the azimuths of the observed bodies is less than about 60°, the use of bisectors is not recommended.

The principal and most useful feature of a bisector is that, by using it as a position line, systematic error is eliminated entirely. A second important feature is that random errors are averaged. This is readily seen from the formula:

$$e = \frac{e_x \sim e_y}{2} \cdot \frac{1}{\sin \theta}$$

It follows that a more reliable position is possible by crossing two bisectors than by crossing three or four position lines.

b. The Three-star Fix

The position lines obtained from astronomical observations of more than two bodies seldom intersect at a common point. Because of errors in observation, computing or plotting, three astronomical position lines usually intersect to form a cocked hat.

By taking pairs of position lines obtained from simultaneous observations of three stars, three bisectors may be drawn. These three bisectors will ALWAYS intersect at a common point. This follows from the simple geometry of the problem.

The most likely reason why three astronomical position lines do not intersect at a common point is that the altitudes obtained from observation are incorrect. This leads to the displacement of one or more of the three position lines from the ship's true position. This, in turn, results in the formation of a cocked hat. In general, the bigger is the cocked hat the bigger is the error in the position lines.

In Fig. 10, P represents a position used to compute intercepts I_A, I_B and I_C from observations of three stars *A, *B and *C whose azimuths differ by about 120° respectively. The resulting position lines AA_1, BB_1 and CC_1 intersect to form a cocked hat.

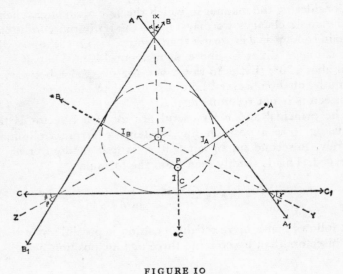

FIGURE 10

If the same systematic error has affected each of the three position lines the ship's true position will lie on each of the three bisectors X, Y and Z. These bisectors intersect at T which is the centre of the in-circle of the cocked hat.

A common method of dealing with a cocked hat is to apply a trial-and-error method by moving each of the position lines (in the imagination) through the same distance either all towards or all away from the directions of the respective observed bodies, until they intersect at a common point which is taken to be the ship's probable position. Such a fix is sometimes called a *cartwheel fix*. The principle employed in this method is the same as that used in the bisector method.

In the example illustrated in Fig. 10, the three intercepts are named towards. Had they been of different senses, or had they been all three away, the ship's true position would have been inside the cocked hat. This will always apply when the systematic error affects each of the position lines, and the observed bodies are spaced equally, or nearly so, around the observer.

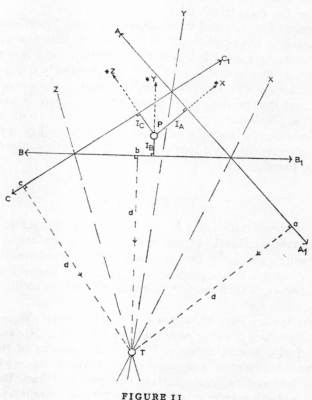

FIGURE 11

Fig. 11 serves to demonstrate that if all the sights are taken on one side of the observer within a sector of 180° or less, the ship's true position will lie outside the cocked hat when the same systematic error affects all sights.

In Fig. 11, P is a position used to compute the three intercepts I_A, I_B and I_C. The three resulting position lines form a cocked hat. The three bisectors intersect at T which is the ship's true position if the same systematic error has affected each of the three position lines.

The point T could have been found by the cartwheel principle. In this case each position line would have been moved through a distance d as illustrated in Fig. 11. It follows that T is the centre of the circle which touches the three position lines at a, b and c, respectively.

A cocked hat provides information that error has affected the sights, but it gives no indication whatever that the error is systematic or random. It is true that if the navigator is sure that systematic error exists he may fix his ship reliably using bisectors or the cartwheel method but, in practice, the navigator is never sure that random error is not present. There is a way by which random error may be detected and, in this event, a navigator may be able to estimate the degree of reliability of his fix. We shall now discuss a four-star fix by which this information may be found.

c. The Four-star Fix

From a consideration of the properties of bisectors it would appear that the ideal requirements for fixing by astronomical observations consist of four position lines obtained from simultaneous sights of four stars equally spaced in azimuth around the observer. The following examples should be studied carefully.

Referring to Fig. 12, suppose that four stars *W, *X, *Y and *Z produce the four position lines AA_1, BB_1, CC_1 and DD_1 respectively. It will be noticed that the azimuths of the four stars are each directed away from the centre of the square formed by the intersection of the intersecting position lines. By using the cartwheel principle, or the bisectors of parallel pairs of position lines, the ship's probable position is at F, the centre of the square.

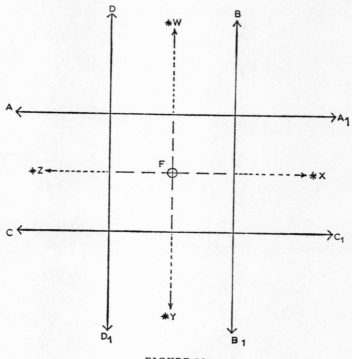

FIGURE 12

The pairs of position lines AA_1 and CC_1, and BB_1 and DD_1, are separated by the same distance. This, coupled with the fact that, relative to F, each of the position lines lies towards the direction of the observed body which produced it, suggests that a common systematic error has affected all four observations. In other words, if four position lines cross as they do in Fig. 12, there is every possibility that a systematic error only has affected the sights and the ship's probable position at the intersection of the bisectors, as illustrated, is a reliable fix.

Had the four position lines obtained from the observations

20

of four stars *W, *X, *Y and *Z intersected as illustrated in Fig. 13, the fix at the intersection of the bisectors would not be such a reliable position as is the case illustrated in Fig. 12.

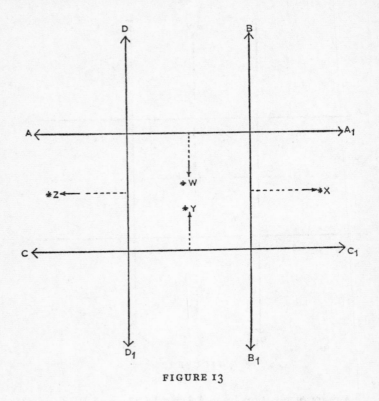

FIGURE 13

It will be noticed in Fig. 13 that by using the cartwheel principle, the effect of moving each position line a given amount in, or away from, the direction of the appropriate observed body, the shape of the area cut off by the intersecting position lines becomes rectangular but not square. This suggests that random error (in addition perhaps to systematic error) has affected the sights.

It follows that by using four position lines which intersect at right angles or nearly so, the navigator may be provided with evidence of the existence in his sights of random error. By using bisectors systematic error will automatically be eliminated, and this sort of error, therefore, need not worry the navigator unduly.

Captain Bini, in his paper, points out that a navigator who has a positive personal error will, when using four position lines in the way described, normally find that the azimuths point away from the centre of the area of intersection of the position lines. The reverse will be the case if personal error is negative. Knowledge of one's personal error, therefore, may be put to good use in assessing the reliability of a four-star fix. Conversely, using four-star fixes of the type described, an effective way of ascertaining one's personal error is provided.

When taking star sights, in order to remove or reduce the possibility of faulty sights, a series of three or five shots of each of the observed stars should be taken in quick succession. After first checking the differences between successive observed altitudes and chronometer times, so that a faulty sight may be detected, the results should be averaged. The average values should then be used in the sight reduction. A fix obtained from a series of single shots should not be regarded as favouring the production of a reliable ship's position and, in general, analysis of position lines obtained in this way is not regarded as being a fruitful procedure.

Appendices

Trigonometry in Nautical Astronomy

The mathematics of nautical astronomy is concerned essentially with the solution of triangles. The branch of mathematics which deals with the computation of unknown parts of triangles is called trigonometry.

Every triangle contains six parts three of which are sides and three angles. If three of the six parts of a plane triangle are given the others may be computed provided that at least one of the given parts is a side. To facilitate the computation of the unknown parts of a triangle trigonometrical functions are used. The principal trigonometrical functions are the six unique ratios of the pairs of sides of a right-angled triangle containing the angle. These ratios are named the *sine* and its reciprocal the *cosecant*; the *secant* and its reciprocal the *cosine*; and the *tangent* and its reciprocal the *cotangent*.

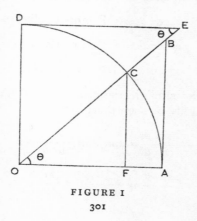

FIGURE I

Tables of trigonometrical functions, natural and logarithmic, are to be found in most collections of nautical tables.

The trigonometrical ratios of any angle θ (less than 90°) may be represented by straight lines as illustrated in Fig. 1.

Let the straight line OA in Fig. 1, of any unit length, be rotated about O so that the angle θ is swept out. Let AOD be 90°. AB and DE are tangents to the circle formed by rotating OA about O:

$$\text{sine } \theta = CF/OC$$

$$\text{secant } \theta = OB/OA$$

$$\text{tangent } \theta = AB/OA$$

But OC and OA are each equal to one unit, therefore:

$$\sin \theta = CF \text{ numerically}$$

$$\sec \theta = OB \text{ numerically}$$

$$\tan \theta = AB \text{ numerically}$$

Also:

$$\text{cosine } \theta = OF/OC$$

$$\text{cosecant } \theta = OE/OD$$

$$\text{cotangent } \theta = DE/OD$$

But OC and OD are each of unit length, therefore:

$$\cos \theta = OF \text{ numerically}$$

$$\text{cosec } \theta = OE \text{ numerically}$$

$$\text{cotan } \theta = DE \text{ numerically}$$

Considering the similar triangles OCF, OBA and ODE, in Fig. 1, it will readily be seen that:

$$\frac{\sin \theta}{\cos \theta} = \tan \theta = \frac{1}{\cot \theta}$$

$$\frac{\sin \theta}{1} = \frac{1}{\text{cosec } \theta} = \frac{\tan \theta}{\sec \theta}$$

$$\frac{\cos \theta}{\sin \theta} = \cot \theta = \frac{1}{\tan \theta}$$

$$\frac{\cos \theta}{1} = \frac{1}{\sec \theta} = \frac{\cot \theta}{\operatorname{cosec} \theta}$$

Also, by Pythagoras' theorem:

$$\sin^2 \theta + \cos^2 \theta = 1$$

$$\tan^2 \theta + 1 = \sec^2 \theta$$

$$1 + \cot^2 \theta = \operatorname{cosec}^2 \theta$$

The sine of an arc may be defined as the ratio between the length of the perpendicular dropped from one extremity of the arc on to the diameter through the other extremity, and the radius itself. The sign which this perpendicular has for all angles up to 180° is regarded as being positive. For all angles more than 180° and less than 360° the perpendicular is regarded as being negative. It follows that the sines of angles between 0° and 180° are positive, and that those of angles between 180° and 360° are negative.

The result of graphing the sine of an angle against angle from 0° to 360°, is a sine curve. From the sine curve it may readily be seen that when θ is in the second quadrant, that is to say, when θ lies between 90° and 180°:

$$\sin \theta = \sin (180 - \theta)$$

Similarly, when θ is in the third quadrant:

$$\sin \theta = -\sin (\theta - 180)$$

When θ is in the fourth quadrant:

$$\sin \theta = -\sin (360 - \theta)$$

A cosine curve has the same shape as a sine curve, but it is 90° out of step, the cosine curve leading the sine curve as illustrated in Fig. 2.

Because, as we have seen, all trigonometrical ratios are functions of the sine and/or cosine, the signs of trigonometrical functions of angles in the second, third and fourth quadrants may readily be ascertained.

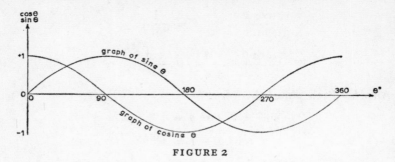

FIGURE 2

In practical navigation angles of more than 180° are not generally considered, so that the signs of trigonometrical functions of angles in the third and fourth quadrants need not concern us. Sines and cosecants of angles in the second quadrant are positive, and cosines, secants, tangents and cotangents are negative.

Practical nautical astronomy involves solving triangles. Plane trigonometry is employed to solve plane triangles which, if right-angled, are readily solved using the trigonometrical ratios direct. Now although every oblique triangle may be split into two right-angled triangles and its solution, therefore, obtained by using right-angled trigonometry, it is necessary on occasions to use the formulae of oblique trigonometry. The more important of these are the sine and cosine formulae.

THE PLANE SINE FORMULA

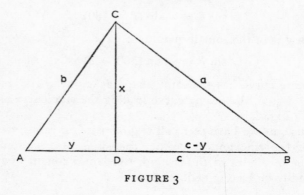

FIGURE 3

Fig. 3 illustrates any plane oblique triangle. The sine formula applied to this triangle is:

$$\frac{a}{\sin A} = \frac{b}{\sin B} = \frac{c}{\sin C}$$

This may be proved as follows:

CD is a perpendicular from C dropped on to AB. Let the length of CD be denoted by x, then:

$$x = b \sin A$$

also

$$x = a \sin B$$

therefore:

$$\frac{a}{\sin A} = \frac{b}{\sin B}$$

By dropping a perpendicular from B on to AC, or from A on to BC, it may be proved in a similar way that:

$$\frac{a}{\sin A} = \frac{c}{\sin C}$$

$$\frac{b}{\sin B} = \frac{c}{\sin C}$$

therefore:

$$\frac{a}{\sin A} = \frac{b}{\sin B} = \frac{c}{\sin C}$$

THE PLANE COSINE FORMULA

Referring to Fig. 3, the cosine formula applied to the triangle ABC is:

$$\cos A = \frac{b^2 + c^2 - a^2}{2bc}$$

from which:

$$a^2 = b^2 + c^2 - 2bc \cos A$$

This may be proved with reference to Fig. 3 as follows:

By applying Pythagoras' theorem to the two right-angled triangles ACD and BCD we have:

$$b^2 = x^2 + y^2 \tag{1}$$

$$a^2 = x^2 + (c - y)^2 \tag{2}$$

Subtracting (2) from (1):

$$b^2 - a^2 = y^2 - (c - y)^2$$

i.e. $$b^2 - a^2 = y^2 - (c^2 + y^2 - 2cy)$$

i.e. $$b^2 - a^2 = 2cy - c^2$$

But $$y = b \cos A$$

therefore:

$$a^2 = b^2 + c^2 - 2bc \cos A$$

or $$\cos A = \frac{b^2 + c^2 - a^2}{2bc}$$

The following trigonometrical identities are connected with compound angles:

$$\sin (A + B) = \sin A \cos B + \cos A \sin B$$

$$\cos (A + B) = \cos A \cos B - \sin A \sin B$$

$$\sin (A - B) = \sin A \cos B - \cos A \sin B$$

$$\cos (A - B) = \cos A \cos B + \sin A \sin B$$

These identities may be proved as follows.

Referring to Fig. 4, let the points P and Q have co-ordinates $(r \cos B, r \sin B)$ and $(r \cos A, r \sin A)$ relative to the axes of reference OX and OY.

By Pythagoras' theorem:

$$PQ^2 = (r \cos A - r \cos B)^2 + (r \sin A - r \sin B)^2 \tag{1}$$

By the plane cosine formula:

$$PQ^2 = r^2 + r^2 - 2r^2 \cos (A - B) \tag{2}$$

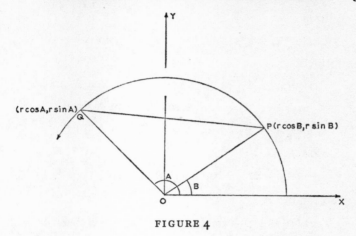

FIGURE 4

Equating the values for PQ² in (1) and (2) we have:

$$2r^2 - 2r^2 \cos(A - B)$$
$$= r^2 \cos^2 A + r^2 \cos B - 2r^2 \cos A \cos B$$
$$+ r^2 \sin^2 A + r^2 \sin^2 B - 2r^2 \sin A \cos B$$

from which:

$$\cos(A - B) = \cos A \cos B + \sin A \sin B$$

By putting B = −B in this identity we have:

$$\cos(A + B) = \cos A \cos B - \sin A \sin B$$

By putting A = (90 − A) we have:

$$\sin(A - B) = \sin A \cos B + \cos A \sin B$$

By putting B = −B in this identity we have:

$$\sin(A + B) = \sin A \cos B + \cos A \sin B$$

Products as Sums and Differences

$$\sin(A + B) = \sin A \cos B + \cos A \sin B$$
$$\sin(A - B) = \sin A \cos B - \cos A \sin B$$

By addition:

$$\sin (A + B) + \sin (A - B) = 2 \sin A \cos B$$

By subtraction:

$$\sin (A + B) - \sin (A - B) = 2 \cos A \sin B$$

$$\cos (A + B) = \cos A \cos B - \sin A \sin B$$

$$\cos (A - B) = \cos A \cos B + \sin A \sin B$$

By addition:

$$\cos (A + B) + \cos (A - B) = 2 \cos A \cos B$$

By subtraction:

$$\cos (A + B) - \cos (A - B) = - 2 \sin A \sin B$$

Sums and Differences as Products

Let

$$X = \tfrac{1}{2}(X + Y) + \tfrac{1}{2}(X - Y)$$

and

$$Y = \tfrac{1}{2}(X + Y) - \tfrac{1}{2}(X - Y)$$

then:

$$\sin X + \sin Y = \sin \tfrac{1}{2}(X + Y) \cos \tfrac{1}{2}(X - Y)$$
$$+ \cos \tfrac{1}{2}(X + Y) \sin \tfrac{1}{2}(X - Y)$$
$$+ \sin \tfrac{1}{2}(X + Y) \cos \tfrac{1}{2}(X - Y)$$
$$- \cos \tfrac{1}{2}(X + Y) \cos \tfrac{1}{2}(X - Y)$$

i.e.

$$\sin X + \sin Y = 2 \sin \tfrac{1}{2}(X + Y) \cos \tfrac{1}{2}(X - Y)$$

Similarly:

$$\sin X - \sin Y = 2 \cos \tfrac{1}{2}(X + Y) \sin \tfrac{1}{2}(X - Y)$$

$$\cos X + \cos Y = 2 \cos \tfrac{1}{2}(X + Y) \cos \tfrac{1}{2}(X - Y)$$

$$\cos X - \cos Y = - 2 \sin \tfrac{1}{2}(X + Y) \sin \tfrac{1}{2}(X - Y)$$

Functions of Small Angles

Remembering that the area of a plane triangle is equal to half the product of the base and the perpendicular height measured

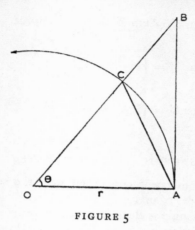

FIGURE 5

from the base; and that the area of a sector is equal to half the product of the radius squared and the angle in radians: referring to Fig. 5 we have:

Area triangle OAB > area sector OAC > area triangle OAC
Therefore:

$$\tfrac{1}{2}r^2 \tan \theta > \tfrac{1}{2}r^2\theta^c > \tfrac{1}{2}r^2 \sin \theta$$

and

$$\tan \theta > \theta^c > \sin \theta$$

and

$$\frac{\tan \theta}{\theta} > 1 > \frac{\sin \theta}{\theta}$$

As θ tends to zero, $(\tan \theta)/\theta$ tends to 1 and $(\sin \theta)/\theta$ tends to 1.
For small angles, therefore, we may write θ^c for either $\sin \theta$ or $\tan \theta$.

Now

$$\cos \theta = \cos (\tfrac{1}{2}\theta + \tfrac{1}{2}\theta)$$

$$= 1 - 2 \sin^2 \tfrac{1}{2}\theta$$

Thus, when θ is small, we may write $(1 - \theta^2/2)$ for $\cos \theta$.
Now 1 radian = 3438′ approx., so that

$$\sin 1' = \tan 1' = \frac{1}{3438} \quad \text{approx.}$$

and $$\sin \theta' = \tan \theta' = \frac{\theta}{3438} \quad \text{approx.}$$

Also $$\cos 1' = \left(1 - \frac{1^2}{2 \times 3438^2}\right)$$

and $$\cos \theta' = \left(1 - \frac{\theta^2}{2 \times 3438^2}\right)$$

SPHERICAL TRIGONOMETRY

Spherical trigonometry is concerned with the several methods of solving spherical triangles.

A spherical triangle is formed on a sphere by the intersection of three *great circles*: a great circle is a circle on the sphere's surface on the plane of which the centre of the sphere lies.

Two great-circle arcs intersect to form a *spherical angle*, the magnitude of which is equivalent to the plane angle between the tangents to the great circle arcs at the point of intersection.

The measure of a spherical arc or side of a spherical triangle is equivalent to the angle at the centre of the sphere contained between the radii which terminate at the ends of the arc.

Every spherical triangle has six parts, three of which are angles, and the other three sides. It is conventional to denote an angle of a spherical triangle by a capital letter, and a side by a small letter corresponding to the letter used for the opposite angle. Thus, if an angle is denoted by X, the side opposite is denoted by x.

If three parts of any spherical triangle are known it is possible to compute any of the other parts direct by means of one of three fundamental formulae. These are the *spherical sine*, cosine and *four parts formulae*.

The Spherical Sine Formula

In any spherical triangle XYZ:

$$\frac{\sin x}{\sin X} = \frac{\sin y}{\sin Y} = \frac{\sin z}{\sin Z}$$

Proof: Referring to Fig. 6, in which the spherical triangle XYZ is depicted on a sphere whose centre is at O.

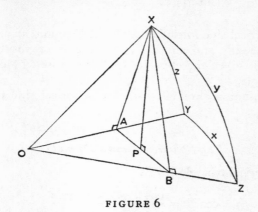

Drop a perpendicular from X on to plane OYZ at P.

Drop perpendiculars from P on to radii OY and OZ at A and B respectively.

Because XA and XB lie in the planes of the arcs XY and XZ respectively, it follows that:

$$\text{Plane angle XAP} = \text{spherical angle XYZ}$$

$$\text{Plane angle XBP} = \text{spherical angle XZY}$$

Now:

$$\frac{\sin z}{\sin Z} = \frac{AX/OX}{XP/BX} = \frac{AX\ BX}{OX\ XP}$$

and

$$\frac{\sin y}{\sin Y} = \frac{BX/OX}{XP/AX} = \frac{BX\ AX}{OX\ XP}$$

therefore:

$$\frac{\sin z}{\sin Z} = \frac{\sin y}{\sin Y}$$

By dropping a perpendicular from Y on to the opposite plane OXZ, and proceeding similarly, it may be proved that:

$$\frac{\sin z}{\sin Z} = \frac{\sin x}{\sin X}$$

21+

therefore:

$$\frac{\sin x}{\sin X} = \frac{\sin y}{\sin Y} = \frac{\sin z}{\sin Z}$$

The spherical sine formula may be used to find an angle given the opposite side and another angle and its opposite side; or to find a side given the opposite angle and another side with its opposite angle.

Because $\sin \alpha = \sin (180 - \alpha)$, the spherical sine formula is ambiguous.

The Spherical Cosine Formula

In any spherical triangle XYZ:

$$\cos X = \frac{\cos x - \cos y \cos z}{\sin y \sin z}$$

or $\qquad \cos x = \cos X \sin y \sin z + \cos y \cos z$

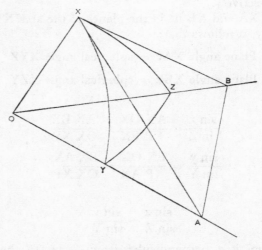

FIGURE 7

Proof: Referring to Fig. 7: Let XYZ be a spherical triangle on the sphere whose centre is at O. Tangents at X drawn in the

planes of the sides XY and XZ meet the plane OYZ at A and B respectively.

Because XA and XB are tangents in the planes of the arcs XY and XZ the plane angle BXA is equal to the spherical angle X. Also OXB and OXA are right angles.

By the plane cosine formula applied to triangles OAB and AXB:

$$AB^2 = OA^2 + OB^2 - 2.OA.OB.\cos x \qquad (1)$$

$$AB^2 = AX^2 + BX^2 - 2.AX.BX.\cos X \qquad (2)$$

Subtract (2) from (1):

$$O = OA^2 + OB^2 - 2.OA.OB.\cos x$$
$$- (AX^2 + BX^2 - 2.AX.BX.\cos X)$$

$$= OA^2 + OB^2 - 2.OA.OB.\cos x$$
$$- AX^2 - BX^2 + 2.AX.BX.\cos X$$

$$= (OA^2 - AX^2) + (OB^2 - BX^2)$$
$$- 2.OA.OB.\cos x + 2.AX.BX.\cos X$$

$$= 2.OX^2 - 2.OA.OB.\cos x + 2.AX.BX.\cos X$$

from which:

$$\cos X = \frac{OA.OB \cos x - OX^2}{AX.BX} \qquad (3)$$

Divide (3) by OA.OB:

$$\cos X = \frac{\cos x - \cos y.\cos z}{\sin y.\sin z}$$

The spherical cosine formula suffers the disadvantage in that it is not suitable for logarithmic computation.

The Four-Parts Formula

In any spherical triangle XYZ, if three of any four adjacent parts are known, the fourth may be found directly by means of the four-parts formula.

In the triangle XYZ depicted in Fig. 8, the four-parts formula connecting angles X and Y and the sides y and z is:

$$\cos z \cos X = \sin z \cot y - \sin X \cot Y$$

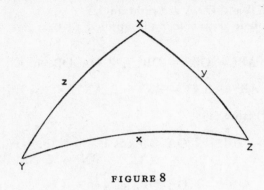

FIGURE 8

Proof: By the spherical cosine formula:

$$\cos y = \cos Y \sin z \sin x + \cos z \cos x \qquad (1)$$

$$\cos x = \cos X \sin y \sin z + \cos y \cos z \qquad (2)$$

By the spherical sine formula:

$$\sin x = \frac{\sin X \sin y}{\sin Y} \qquad (3)$$

Substitute (2) for cos x in (1); and (3) for sin x in (1). Thus:

$$\cos y = \cos Y \sin z \frac{\sin X \sin y}{\sin Y}$$
$$+ \cos z(\cos X \sin y \sin z + \cos y \cos z)$$

$$= \cot Y \sin X \sin y \sin z$$
$$+ \cos z \cos X \sin y \sin z + \cos y \cos^2 z$$

$$\cos y - \cos y \cos^2 z = \sin y \sin z(\cot Y \sin X + \cos z \cos X)$$

$$\cos y(1 - \cos^2 z) = \sin y \sin z(\cot Y \sin X + \cos z \cos X)$$

$$\frac{\cos y \times \sin^2 z}{\sin y \sin z} = \cot Y \sin X + \cos z \cos X$$

or:

$$\cos z \cos X = \sin z \cot y - \sin X \cot Y$$

Napier's Rules of Circular Parts

If one of the angles in a spherical triangle XYZ is 90°, the fundamental formulae reduce to simple expressions, each involving three terms only. This is so because $\sin 90° = 1$, and $\cos 90° = 0$.

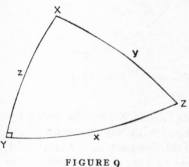

FIGURE 9

In the spherical triangle XYZ depicted in Fig. 9,
Because $Y = 90°$:

$$\sin Z = \sin z \, \text{cosec} \, y$$

$$\cos x = \cos y \cos z$$

$$\cot X = \cot x \sin z$$

It is possible to derive ten such formulae which, collectively, provide the means of solving every case of right-angled triangles. Instead of deducing from these formulae ten distinct rules for the solution of the various cases, the whole, by means of the assistance of an ingenious contrivance invented by the illustrious Baron Napier, may be comprehended in two simple rules known as 'Napier's rules'.

21*

The parts of the right-angled triangle (not including the 90° angle) are written in order in the five sectors of the cartwheel illustrated in Fig. 10.

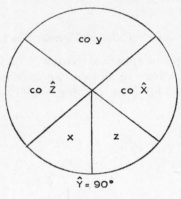

FIGURE 10

The two angles X and Z and the side opposite to the right angle Y, are prefixed with 'co' meaning complement.

Of any three of the five parts in the cartwheel, one is a 'middle' part, and the other two are either 'opposite' or 'adjacent' parts. Napier's mnemonic rules are:

sine middle part = product cosines opposites

sine middle part = product tangents adjacents

Any oblique spherical triangle may be divided into two right-angled triangles by dropping a perpendicular great circle from any apex on to the opposite side or side produced. It follows, therefore, that Napier's simple formulae may be used to solve any oblique triangle indirectly and without resort to the fundamental formulae of spherical trigonometry. They are, therefore, powerful artifices in the practice of navigation, being particularly important in the construction of short-method navigation tables.

In astronomical navigation the more important spherical trigonometrical problems are those in which it is required to find an angle given three sides; or those in which it is required to find a side given the other two sides and the included angle. The spherical cosine formula is, therefore, the basis of the solutions of most nautical astronomical problems.

Because the spherical cosine formula is not suitable for logarithmic computation, other formulae derived from the cosine formula, and which are suitable for use with logarithms, are invariably used by navigators.

The trigonometrical functions versine and haversine are functions used almost exclusively by navigators.

$$\text{versine } \theta = 1 - \cos \theta$$
$$\text{haversine } \theta = \tfrac{1}{2}(1 - \cos \theta)$$

The great value of the versine is that its sign is positive for all angles, so that the various forms of the versine and haversine formulae help to eliminate or reduce the seaman's traditional difficulty of dealing with trigonometrical functions of angles over 90°.

The Calculus and Nautical Astronomy

The calculus is the branch of mathematics in which the operation of *taking a limit* plays a predominant role. A limit of great importance in the calculus is called a *derivative of a function*, and the process of finding it is called *differentiation*.

A *function* in mathematics is a quantity the value of which depends upon the value of some other quantity. The area of a circle, for example, is dependent upon the radius of the circle, so that we say that the area of a circle is a function of its radius. Similarly the sine of an angle is a function of the angle; and the draught of a ship is a function of her displacement, etc, etc.

If a variable quantity is denoted by x, an expression which involves x is a function of x. The expression $3x^2 + 2x$ is a function of x, and so is $(x^2 - 4x + 2)$ a function of x.

If $y = 3x^2 + 2x$ we say that y is a function of x and, in the normal notation, this is written as:

$$y = f(x)$$

The expression $y = 3x^2 + 2x$ may be represented graphically by plotting points relative to two mutually perpendicular axes of reference which are graduated with values of the variables x and y respectively. A curve drawn through a relatively small number of plotted points represents the equation $y = 3x^2 + 2x$.

The positions of the points through which the curve is drawn are found by assigning values to x in the equation and finding the corresponding values of y. This process is familiar to all navigators.

Many of the practical applications of the calculus, and in

particular the applications to nautical astronomy, are related to the idea of the gradient at a point on a curve. The gradient of a curve at any point on it is defined as the slope of a straight line tangential to the curve at the point. An example will make this clear.

EXAMPLE: Find the gradient of the curve $y = 3x^2$ at the point (3, 27).

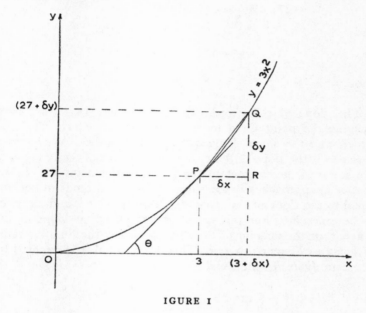

FIGURE 1

The quantities (3, 27) given in the question are the distances respectively x-wards from the y axis and y-wards from the x axis. The first quantity 3, is called the abscissa, and the second quantity 27, is called the ordinate of the point P. The two quantities form the co-ordinates of the point P illustrated in Fig. 1.

Referring to Fig. 1, consider the point Q on the curve, the co-ordinates of which are $(3 + \delta x)$, $(27 + \delta y)$.

The expressions δx and δy are used to denote small quantities, so that $(3 + \delta x)$ may be expressed as 'three plus a little bit x-wards of the y axis', and $(27 + \delta y)$ may be expressed as '27 plus a little bit y-wards of the x axis'.

Now consider the triangle PQR, in which:

$$PR = \delta x$$
$$QR = \delta y$$

and $$PQ = \text{chord of curve } y = 3x^2$$

From the equation of the curve:

$$(27 + \delta y) = 3(3 + \delta x)^2$$

and $$\delta y = 3(9 + 6\delta x + (\delta x)^2) - 27$$

i.e. $$\delta y = 18\delta x + 3(\delta x)^2$$

and $$\frac{\delta y}{\delta x} = 18 + 3\delta x$$

The slope of the chord PQ is, therefore, $18 + 3\delta x$. Now imagine the point Q to move along the curve towards P, in which case the slope changes and δx diminishes. As Q moves closer to P the slope of the chord becomes more nearly equal to the slope of the tangent to the curve at the point P. We say that as Q approaches P, the slope of the chord tends to become equal to the slope of the tangent to the curve at P and δx tends to become zero. For this reason we think of the slope of the tangent, or the tangent of the slope, as being the limiting value of $\delta y/\delta x$ as δx tends to zero. This limiting value is denoted by the term dy/dx. In the usual notation:

$$\tan \theta = \mathop{Lt}_{\delta x \to 0} \frac{\delta y}{\delta x} = \frac{dy}{dx}$$

The quantity dy/dx is called the *derived function of x*, or the *derivative of x*. It is usually denoted by $Df(x)$ or $f'x$.

Let us return to our example. We have seen that if $y = 3x^2$, $\delta y/\delta x = 18 + 3\delta x$. It now remains to be seen how dy/dx is found from this expression.

Remembering that dy/dx is the limiting value of $\delta y/\delta x$ as δx tends to zero, our problem is to find the limiting value of $18 + 3\delta x$ as δx tends to zero. This limiting value is 18, so that if $y = 3x^2$, the gradient of the curve at the point (3, 27) is 18. In other words the tangent to the curve at the point (3,27) makes an angle with the x axis equal to $\tan^{-1} 18$.

The process of finding a derived function amounts to the evaluation of a limit. In the general case of the curve $y = f(x)$ the gradient at any point where $x = c$ is:

$$\underset{\delta x \to 0}{Lt} \frac{\delta y}{\delta x} = \underset{\delta x \to 0}{Lt} \frac{f(x + \delta x) - f(c)}{\delta x}$$

EXAMPLE: If $y = x^4$ find dy/dx.

$$Df(x) = Dx^4$$

$$= \underset{\delta x \to 0}{Lt} \frac{(x + \delta x)^4 - x^4}{\delta x}$$

$$= \underset{\delta x \to 0}{Lt} \frac{4x^3\delta x + 6x^2(\delta x)^2 + 4x(\delta x)^3 + (\delta x)^4}{\delta x}$$

$$= \underset{\delta x \to 0}{Lt} \, 4x^3 + 6x^2\delta x + 4x(\delta x)^2 + (\delta x)^3$$

$$= 4x^3$$

It may be proved that if n is any number, positive, negative or fractional; then, if $y = x^n$,

$$\frac{dy}{dx} = nx^{(n-1)}$$

and if $y = ax^n$,

$$\frac{dy}{dx} = nax^{(n-1)}$$

dy and dx are called *differentials*, and dy/dx is called the *differential coefficient of y with respect to x*.

The results of differentiating $y = x^n$ or $y = ax^n$ may be written respectively as $dy = nx^{(n-1)} \, dx$ and $dy = nax^{(n-1)} \, dx$.

To Differentiate a Sum

Let $y = U_1 + U_2 + U_3 + \ldots$
where U_1, U_2, U_3, etc., are functions of x.

Let y increase to $(y + \delta y)$, U_1 increase to $(U_1 + \delta U_1)$, U_2 increase to $(U_2 + \delta U_2)$, etc. Then:

$$y + \delta y = U_1 + \delta U_1 + U_2 + \delta U_2 + U_3 + \delta U_3 + \ldots$$

$$\delta y = \delta U_1 + \delta U_2 + \delta U_3 + \cdots$$

and,

$$\frac{\delta y}{\delta x} = \frac{\delta U_1}{\delta x} + \frac{\delta U_2}{\delta x} + \frac{\delta U_3}{\delta x} + \cdots$$

In the limit, as $\delta x \to 0$,

$$\frac{\delta y}{\delta x} \to \frac{dy}{dy}; \quad \frac{\delta U_1}{\delta x} \to \frac{dU_1}{dx}; \quad \frac{\delta U_2}{\delta x} \to \frac{dU_2}{dx}; \cdots$$

Therefore:

$$\frac{dy}{dx} = \frac{dU_1}{dx} + \frac{dU_2}{dx} + \frac{dU_3}{dx} + \cdots$$

Function of a Function

If $y = (2x^3 + 2x)^6$, then

$$y = f(x)^6$$

Let $f(x) = U$, then

$$y = U^6$$

Now y is a function of U and U is a function of x.
Therefore, y is a function of a function of x.
Now:

$$\frac{\delta y}{\delta x} = \frac{\delta y}{\delta U} \times \frac{\delta U}{\delta x} \quad \text{(by the rules of algebra)}$$

In the limit, as $\delta x \to 0$ so also does $\delta U \to 0$, and

$$\frac{\delta y}{\delta x} \to \frac{dy}{dx}$$

and,

$$\frac{\delta y}{\delta U} \to \frac{dy}{dU}$$

and,

$$\frac{\delta U}{\delta x} \to \frac{dU}{dx}$$

Therefore:

$$\frac{dy}{dx} = \frac{dy}{dU} \times \frac{dU}{dx}$$

Thus, in the example given above:

$$\frac{dy}{dx} = 6(2x^3 + 2x)^5 \times (6x^2 + 2)$$

i.e.

$$\frac{dy}{dx} = (36x^2 + 12)(2x^3 + 2x)^5$$

The Differentiating of a Product

Let $y = UV$, where $U = f(x)$ and $V = f(x)$. Then:

$$y + \delta y = (U + \delta U)(V + \delta V)$$
$$= UV + V\,\delta U + U\,\delta V + \delta U\,\delta V$$
$$\delta y = V\,\delta U + U\,\delta V + \delta U\,\delta V$$

and

$$\frac{\delta y}{\delta x} = \frac{V\,\delta U}{\delta x} + \frac{U\,\delta V}{\delta x} + \frac{\delta U\,\delta V}{\delta x}$$

In the limit, as $\delta x \to 0$, $\delta y/\delta x \to dy/dx$, etc., and

$$\frac{dy}{dx} = V\frac{dU}{dx} + U\frac{dV}{dx}$$

The Differentiating of a Quotient

Let $y = U/V$, where $U = f(x)$ and $V = f(x)$. Then

$$y + \delta y = \frac{U + \delta U}{V + \delta V}$$

$$\delta y = \frac{U + \delta U}{V + \delta V} - \frac{U}{V}$$

$$= \frac{V(U + \delta U) - U(V + \delta V)}{V(V + \delta V)}$$

$$= \frac{V\,\delta U - U\,\delta V}{V(V + \delta V)}$$

$$\frac{\delta y}{\delta x} = \frac{V\,\delta U/\delta x - U\,\delta V/\delta x}{V(V + \delta V)/\delta x}$$

22

In the limit, as $\delta x \to 0$, $\delta y/\delta x \to dy/dx$, etc., and

$$\frac{dy}{dx} = \frac{V\,dU/dx - U\,dV/dx}{V^2}$$

Trigonometrical Functions

Let $y = \sin x$. Then

$$y + \delta y = \sin(x + \delta x)$$

$$\delta y = \sin(x + \delta x) - \sin x$$

$$= 2 \cos(x + \delta x/2) \sin \delta x/2$$

$$\frac{\delta y}{\delta x} = \frac{2 \cos(x + \delta x/2) \sin \delta x/2}{\delta x}$$

$$= \cos(x + \delta x/2) \frac{\sin \delta x/2}{\delta x/2}$$

In the limit, as $\delta x \to 0$, $\delta y/\delta x \to dy/dx$,

$$\cos(x + \delta x/2) \to \cos x$$

$$\frac{\sin \delta x/2}{\delta x/2} \to 1$$

Therefore:

$$\frac{dy}{dy} = \cos x$$

or

$$dy = \cos x\,dx$$

Let $y = \cos x$. Then

$$y + \delta y = \cos(x + \delta x)$$

$$\delta y = \cos(x + \delta x) - \cos x$$

$$= -2 \sin(x + \delta x/2) \sin \delta x/2$$

$$\frac{\delta y}{\delta x} = -\sin(x + \delta x/2) \frac{\sin \delta x/2}{\delta x/2}$$

In the limit, as $\delta x \to 0$, $\delta y / \delta x \to dy/dx$,

$$-\sin (x + \delta x/2) = -\sin x$$

$$\frac{\sin \delta x/2}{\delta x/2} \to 1$$

Therefore: $$\frac{dy}{dx} = -\sin x$$

or, $$dy = -\sin x dx$$

By using the quotient or product rule, and remembering that $\tan \theta = \sin \theta / \cos \theta$, $\cot \theta = \cos \theta / \sin \theta$, $\sec \theta = 1/\cos \theta$, and $\operatorname{cosec} \theta = 1/\sin \theta$, it may readily be shown that:

If $y = \tan \theta$, $\quad dy/d\theta = \sec^2 \theta$

If $y = \cot \theta$, $\quad dy/d\theta = \operatorname{cosec}^2 \theta$

If $y = \sec \theta$, $\quad dy/d\theta = \sec \theta \tan \theta$

If $y = \operatorname{cosec} \theta$, $\quad dy/d\theta = -\operatorname{cosec} \theta \cot \theta$

The differential calculus is of great use in nautical astronomy in connection with small errors in altitude or time and their effects on position lines, and when dealing with rates of change.

To find the rate of change of altitude or azimuth of a celestial body the problem is, in essence, the same as finding the tangent to a curve which connects altitude or azimuth with hour angle. Let us see how the spherical cosine formula applied to the PZX triangle may be dealt with when finding, say, the rate of change of a celestial body's altitude with time assuming the latitude of the observer and the declination of the body to be constant.

$$\cos ZX = \cos P \sin PZ \sin PX + \cos PZ \cos PX$$

i.e. $$\cos z = \cos h \sin PZ \sin PX + \cos PZ \cos PX$$

Now the differential of $\cos z$ (i.e. $d(\cos z)$) is $-\sin z\, dz$ and the differential of $\cos h$ $\sin PZ \sin PX + \cos PZ \cos PX$ is $-\sin h\, dh \sin PZ \sin PX$. Therefore:

$$-\sin z\, dz = -\sin h\, dh \sin PZ \sin PX$$

and the rate of change of z with respect to h is:

$$\frac{dz}{dh} = \sin h \sin PZ \sin PX \operatorname{cosec} z$$

By the sine rule of spherical trigonometry:

$$\sin Z = \sin PX \sin h \operatorname{cosec} z$$

Therefore, by substitution:

$$\frac{dz}{dh} = \sin PZ \sin Z$$

i.e.

$$\frac{dz}{dh} = \cos \phi \sin Z$$

This rate is $(15 \cos \phi \sin Z)'$ per minute.

If δz is regarded as being a small error in altitude due to a small error δh in time, we have:

$$\frac{\delta z}{\delta h} = \cos \phi \sin Z$$

or:

Error in altitude (δz) = Error in time $(\delta h) \cos \phi \sin Z$

Bibliographical Note

Anderson, E. W. *The Principles of Navigation*. London, 1966. A very complete treatise dealing essentially with the philosophy of navigation.

Anon. *Admiralty Manual of Navigation*. 3 volumes. H.M.S.O. London, 1955. The standard work on navigation as practised in the Royal Navy.

Bowditch, N. (originally). *American Practical Navigator*. U.S.N. Hydrographic Office, 1959. A magnificent volume designed specifically for the U.S. Navy.

Cotter, C. H. *The Astronomical and Mathematical Foundations of Geography*. London, 1966. Useful background reading for navigators.

———. *A History of Nautical Astronomy*. London, 1968. Traces the development of astronomical navigation from the times of the earliest sea voyages to the present day.

Dutton, B. (originally). *Navigation and Piloting*. U.S. Naval Institute, 1958. A useful manual of practical navigation.

Harboard, J. B. *Glossary of Navigation*. 4th edition by C. W. T. Layton. Glasgow, 1938. A navigator's *vade mecum*.

Hewson, J. B. *A History of the Practice of Navigation*. Glasgow, 1951. A well-produced volume dealing generally with the history of navigation.

Lecky, S. T. L. *Wrinkles in Practical Navigation*. 9th, and last, edition by the author, 1894. A very large edition, and copies are often to be seen in second-hand bookshops at a moderate price. Latest edition by G. Cobb, London, 1956.

Raper, H. *The Practice of Navigation*. 20th edition by W. Hall and H. B. Goodwin. London, 1914. The 19th-century classic on the subject.

Taylor, E. G. R. *The Haven Finding Art*. London, 1956. An illuminating study of the history of navigation for the general reader, and one which every navigator interested in his subject will find enjoyable reading.

Weems, P. and Lee, C. V. *Marine Navigation*. Annapolis, 1958. A fine text on navigation, whose principal author is one of the outstanding authorities on navigation of the present time.

Serious students of navigation derive inestimable value from the quarterly journals of the Institute of Navigation, London, and the Institute of Navigation, Washington, D.C., U.S.A. These journals are readily available to members of the institutes.

Index

INDEX